Inventing Pollution

Ohio University Press Series in Ecology and History

James L. A. Webb, Jr., Series Editor

Inventing Pollution

Coal, Smoke, and Culture in Britain since 1800

Peter Thorsheim

With a New Preface by the Author

OHIO UNIVERSITY PRESS

ATHENS

Ohio University Press, Athens, Ohio 45701
© 2006, 2017 by Ohio University Press
www.ohiou.edu/oupress/
Printed in the United States of America

28 27 26 25 24 23 22 21 20 19 18   5 4 3 2 1

Cover art: Frederick H. Evans (British, 1853–1943), London. Embankment., 1908, Lantern slide 4.2 × 6.4 cm (1 11/16 × 2 9/16 in.), The J. Paul Getty Museum, Los Angeles

ISBN for 2018 reprint: 978-0-8214-2311-0

Library of Congress Cataloging-in-Publication Data
Thorsheim, Peter.
Inventing pollution : coal, smoke, and culture in Britain since 1800 / Peter Thorsheim.
p. cm. — (Ohio University Press series in ecology and history)
Includes bibliographical references and index.
ISBN 0-8214-1680-4 (cloth : alk. paper) — ISBN 0-8214-1681-2 (pbk. : alk. paper)
1. Air—Pollution—Great Britain—History. 2. Smoke prevention—Great Britain—History. 3. Environmentalism—Great Britain—History. 4. Air—Pollution—Social aspects—Great Britain—History. 5. Coal—Combustion—Health aspects—Great Britain—History. I. Title. II. Series.
TD883.7.G7T48 2006
363.739'20941—dc22
2005029428

For Gina, Erik, and Jacob

Contents

Illustrations

Preface

IT'S EASY TO assume that environmental problems are entirely recent in origin, but that's far from the case. Although many of the hazards that threaten human health and the natural world—such as nuclear wastes, synthetic pesticides, and petrochemicals—are quintessentially modern, other types of pollution have a much longer history. This book explores humanity's long and complicated interaction with coal, the energy source that fueled the Industrial Revolution and the main factor driving climate change. As people in the world's first industrial nation experienced, came to understand, and tried to solve the myriad problems that accompanied their Faustian bargain with this fossil fuel, they created environmental pressure groups, launched scientific efforts to measure and study emissions, enacted pollution regulations, and sought ways to reduce and mitigate the unintended consequences of technological change. In short, they invented the idea of environmental pollution.

The ancients were well aware of the flammable properties of the dark sedimentary rock known as coal, but they viewed it largely as a curiosity. For centuries, its use remained extremely limited, both quantitatively and geographically. To generate heat, people throughout most of the world found it far easier to use wood as a fuel. Britain, however, was an early exception. Starting in the thirteenth century, population growth and deforestation caused shortages of firewood, and Londoners began to adopt coal as their principal source of heat.[1] New uses for coal

In addition to those I thank in the acknowledgments, I wish to express my appreciation to all who have offered questions and comments about this book since its publication—particularly my students at the University of North Carolina at Charlotte.

1. Peter Brimblecombe, *The Big Smoke: A History of Air Pollution in London since Medieval Times* (London: Methuen, 1987); William M. Cavert, *Smoke of London: Energy and Environment in the Early Modern City* (Cambridge: Cambridge University Press, 2017).

emerged in the late eighteenth century, following improvements to the steam engine by the Scottish inventor James Watt. For the first time in history, industry and transportation could be powered not by muscles, wind, and flowing water, but by machines that converted the heat of combustion into mechanical energy.

Coal generated great wealth for industrialists and provided employment for millions of people in mines, factories, and the transport sector. A miracle substance in many respects, it also brought enormous problems, many of which were immediately apparent. At first, the extraction of coal involved nothing more elaborate than breaking it free from surrounding rock in veins that emerged on the earth's surface. Over time, miners followed coal deposits deeper and deeper underground, a development aided by the steam engine, which was used to pump groundwater out of mines to prevent them from flooding.[2] Deep mines were extremely dangerous places to work. As coal mining became a major industry in Britain and other countries during the nineteenth century, thousands died each year in explosions and other accidents, and occupational exposure to coal dust caused debilitating and often fatal harm to many of those fortunate enough to have avoided violent death underground.[3]

Mining disasters and the risks of black lung haunted miners and their families for generations, but these dangers affected a relatively small percentage of the populace. A far more significant hazard than coal mining, in terms of the numbers it affected, was coal combustion. Wherever people burned coal, they filled the air with both visible and invisible pollutants. Scientific identification of the constituents of coal smoke emerged slowly, and medical understanding of their effects took even longer. Today we know that coal emissions include fine particles of soot and ash, as well as carbon dioxide, sulphur, a host of highly toxic

---

2. Rolf Peter Sieferle and Michael P. Osmann, *The Subterranean Forest: Energy Systems and the Industrial Revolution* (Winwick, UK: White Horse Press, 2010).

3. Barbara Freese, *Coal: A Human History* (Cambridge, Mass.: Perseus, 2003); Thomas G. Andrews, *Killing for Coal: America's Deadliest Labor War* (Cambridge, Mass.: Harvard University Press, 2008).

Figure 0.1. Global carbon emissions from fossil fuels, 1900–2014. Source: T. A. Boden, G. Marland, and R. J. Andres (2017), *Global, Regional, and National Fossil-Fuel CO2 Emissions.* Carbon Dioxide Information Analysis Center, Oak Ridge National Laboratory, U.S. Department of Energy, Oak Ridge, Tenn. doi 10.3334/CDIAC/00001_V2017.

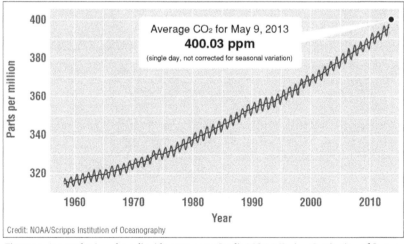

Figure 0.2. Atmospheric carbon dioxide, 1958–2013. Credit: NOAA/Scripps Institution of Oceanography. Source: NASA, *Global Climate Change: Vital Signs of the Planet,* accessed 17 Sept. 2017, https://climate.nasa.gov/system/image_uploads/main/graph-co2-1200x720.png.

organic compounds, and poisonous metals such as arsenic and mercury. Exposure to coal smoke increases the risk of heart disease, stroke, respiratory diseases, and cancer.[4]

The last deep coal mine in Britain ceased operation in 2015, and the remaining surface mines produce even less today than what British miners extracted in 1800.[5] Long a major exporter of coal, Britain now relies on imports for the modest amount of coal that it continues to burn. The air of London and other large cities is now far clearer than it was throughout much of the nineteenth and twentieth centuries. Yet significant threats to air quality remain in Britain, now caused mainly by the millions of diesel and gasoline vehicles that crowd the country's cities. Air pollution in London often exceeds regulatory limits, and the government recently announced that air pollution causes between 40,000 and 50,000 early deaths in the United Kingdom every year.[6]

At the same time that coal production has declined drastically in Britain, it has expanded sharply in other countries. Worldwide coal production has more than doubled in recent decades, rising from approximately 3,000 million tons in 1973 to 7,700 million tons in 2015.[7] Rapidly expanding fossil fuel consumption, combined with weak and poorly enforced environmental protections in many countries, recently led the World Health Organization to declare air pollution "the world's largest single environmental health risk." Across the globe, the WHO

4. M. Zhou et al., "The Associations between Ambient Air Pollution and Adult Respiratory Mortality in 32 Major Chinese Cities, 2006–2010," *Environmental Research* 137 (2015): 278–86; S. Buchanan, E. Burt, and P. Orris, "Beyond Black Lung: Scientific Evidence of Health Effects from Coal Use in Electricity Generation," *Journal of Public Health Policy* 35 (2014): 266–77.

5. Department for Business, Energy & Industrial Strategy, "Digest of UK Energy Statistics (DUKES)," *UK National Statistics*, last updated 27 July 2017, https://www.gov.uk/government/statistics/solid-fuels-and-derived-gases-chapter-2-digest-of-united-kingdom-energy-statistics-dukes.

6. "MPs: UK Air Pollution Is a 'Public Health Emergency,'" *Guardian*, 26 Apr. 2016, https://www.theguardian.com/environment/2016/apr/27/uk-air-pollution-public-health-emergency-crisis-diesel-cars.

7. "Key World Energy Statistics," *International Energy Agency*, accessed 15 Sept. 2017, http://www.iea.org/publications/freepublications/publication/KeyWorld2016.pdf.

estimates that air pollution killed seven million people in 2012 alone.[8] The problem is greatest in countries that are following the pattern that Britain and the world's other wealthiest nations pioneered in the nineteenth century: rapid industrial development, combined with weak or nonexistent environmental protections.

Today, the entire world faces the grave threat of human-induced climate change, caused mainly by the billions of tons of carbon dioxide we have poured into the atmosphere through the burning of coal and other fossil fuels since the start of the Industrial Revolution. For more than half a century, scientists have conducted precise measurements of the constituents of the earth's atmosphere, which have revealed a clear and disturbing trend: each year, the concentration of carbon dioxide is higher than the previous year. Most of the increase in carbon dioxide comes from coal and oil. All fossil fuels contain carbon, but some have more of this element than others. Natural gas and liquid fuels derived from petroleum contain much more hydrogen than does coal, so their combustion emits less carbon dioxide than does an equivalent quantity of coal. The type of coal that yields the most carbon dioxide in relation to its energy content is anthracite, paradoxically praised by smoke-abatement advocates because it produces less visible smoke than other solid fossil fuels, such as bituminous coal or lignite. Anthracite produces nearly twice as much carbon dioxide as natural gas in generating the same amount of heat.[9]

As hard as they tried, the early environmentalists who rallied behind the cause of smoke abatement succeeded neither technologically nor politically in ridding the air of pollution. Yet they did not strive in vain. In many ways, they faced a hurdle similar to the one we now face with climate change: convincing those in positions of economic and political power of the seriousness of the problem. In December 2015, representatives of 195 nations unanimously approved the Paris Climate Accord, which for the first time committed each signatory country to

---

8. "7 Million Premature Deaths Annually Linked to Air Pollution," *World Health Organization,* news release, 25 Mar. 2014, http://www.who.int/media-centre/news/releases/2014/air-pollution/en/.

9. "How Much Carbon Dioxide Is Produced When Different Fuels Are Burned?," *U.S. Energy Information Administration,* last reviewed 8 June 2017, https://www.eia.gov/tools/faqs/faq.php?id=73&t=11.

reduce its emissions of greenhouse gases. Crucially, the agreement included rich and poor nations alike, and it gained the support of the two countries that produce the greatest amount of carbon dioxide, China and the United States.[10] The latter have been far slower than many countries to acknowledge the dangers of climate change and their own roles in contributing to it, but in recent years both countries have begun to make important strides through investments in renewable energy, efficiency, and efforts to reduce greenhouse emissions from fossil fuel use. In 2007, the US Supreme Court ruled that the Environmental Protection Agency possessed the authority to limit carbon dioxide from vehicles, and seven years later it asserted that the EPA could regulate emissions from new and updated power plants and factories.[11] This development is yet another chapter in the long and continually unfolding story that began in Britain two centuries ago when a constellation of farsighted individuals first invented the concept of environmental pollution. In light of Donald Trump's hostility toward international cooperation in general and the Paris agreement in particular, grave doubts exist about whether the US and other nations will sustain the commitments they have made to address carbon pollution—and whether these will be sufficient to avert catastrophic climate change.

---

10. Coral Davenport, "Nations Approve Landmark Climate Accord in Paris," *New York Times,* 12 Dec. 2015.

11. "Justices Rule against Bush Administration on Emissions," *New York Times,* 2 Apr. 2007; "Justices Uphold Emission Limits on Big Industry," *New York Times,* 24 June 2014.

Acknowledgments

THIS BOOK COULD NOT HAVE been begun, much less completed, without the inspiration, guidance, and support of a great many people. I owe immense gratitude to the scholar-teachers who shaped my development as a historian, especially Bob Bonner and Diethelm Prowe at Carleton College; James S. Donnelly Jr., William Cronon, Rudy Koshar, Suzanne Desan, Eric Schatzberg, David Lindberg, Johann Sommerville, and the late David Woodward at the University of Wisconsin at Madison; and Colin Jones, Hilary Marland, and the late Joan Lane at the University of Warwick. I have also benefited greatly from conversations with and encouragement from scholars elsewhere, including Peter Brimblecombe, E. Melanie DuPuis, Sarah Elkind, Angela Gugliotta, Christopher Hamlin, J. M. Harrington, Gerry Kearns, Bill Luckin, John R. McNeill, Geneviève Massard-Guilbaud, Martin V. Melosi, Betsy Mendelsohn, Stephen Mosley, Matthew Osborn, John Pickstone, Sara Pritchard, Harold Platt, John Ranlett, Christine Meisner Rosen, Edmund Russell, Christopher Sellers, Joel A. Tarr, Nancy Tomes, Douglas Weiner, Verena Winiwarter, and Anthony Wohl.

My friends and fellow graduate students at Wisconsin and Warwick were an invaluable source of intellectual and emotional camaraderie. I especially want to thank David Burrow, Fa-ti Fan, Robert Goodrich, James M. Lee, Thomas M. Lekan, Timothy McMahon, Michael de Nie, Jared Orsi, Jonathan Reinarz, James A. Spiller, David Stradling, Gregory Summers, Judith Wamser, and Deirdre Weaver. In Madison, the staff of the U.W. History Department, particularly Mary Moffatt and the late Judy Cochran, provided assistance and encouragement throughout my graduate career.

Heartfelt thanks are due to the archivists, librarians, and support staff who assisted me during this project, including Julie Basinger and Kelly Hatley of the Department of History at the University of North Carolina at Charlotte, and library staff at the University of North Carolina at Charlotte, the University of Wisconsin at Madison, Duke University, and Harvard University. On the other side of the Atlantic, I would like to thank Paul Taylor of the Local Studies and History Service in Birmingham,

Barry Neild of British Engine Insurance Ltd., Alistair Tough of Glasgow University Archive Services, Peter Ross of the Guildhall Library, Peter McNiven of the John Rylands University Library of Manchester, Brian Rance of the King Alfred School, Richard Storey and Christine Woodland of the Modern Records Centre at Warwick University, Helen Ford and Kevin Bodily of the National Gas Archive, Mary Stevens of the National Society for Clean Air and Environmental Protection, Janet Hartley of the National Trust, Colin Penman of the Royal Academy of Arts, James Beaton of the Royal College of Physicians and Surgeons of Glasgow, Alastair McCapra of the Royal Society for the Promotion of Health, and Stephen Lowther of the Wellcome Library for the History and Understanding of Medicine. I am particularly grateful to Cerise Lawson-Tancred for her generosity in allowing me to quote from the papers of her father, the late Sir Hugh Beaver.

The research for this book was funded in part by a pre-dissertation fellowship from the Council for European Studies, graduate fellowships from the University of Wisconsin at Madison and the University of Warwick, a Bernadotte E. Schmitt Grant from the American Historical Association, and research funding from the Faculty Grants Committee of the University of North Carolina at Charlotte. I also wish to thank the latter for supporting the printing costs of this book.

For their valuable advice, astute criticism, and generous encouragement as I completed this book, I am grateful to Ross Bassett, Mark Cioc, Jerry Dávila, James K. Hogue, Lyman Johnson, Thomas M. Lekan, John R. McNeill, Martin V. Melosi, Gregory Mixon, Michael de Nie, Alan Rauch, John David Smith, and Joel A. Tarr. Finally, I thank the superb team I worked with at Ohio University Press, especially Gillian Berchowitz, Ricky S. Huard, Ed Vesneske Jr., and James L. A. Webb, Jr.

My greatest appreciation is to my wife, Gina Campbell, for all of her support, encouragement, insightful comments—and fortitude— throughout the years that I have worked on this project.

Timeline

1851 City of London gains authority to levy fines on factory owners for excessive smoke within the "Square Mile"

1853 Smoke Nuisance Abatement (Metropolis Act) grants power to punish smoky industries throughout the Metropolitan Police District

1858 Great Stink afflicts London and prompts new sewer system

1859 Robert Angus Smith identifies acid rain and attributes it to coal combustion

1863 Alkali Act establishes world's first pollution-control agency

1872 National Health Society established in London, with the Duke of Westminster as president

1876 Manchester and Salford Noxious Vapours Abatement Association established

1876 Kyrle Society established in London by Miranda and Octavia Hill

1877 Society for the Protection of Ancient Buildings founded in London by William Morris

1880 Fog and Smoke Committee established in London by Ernest Hart and Octavia Hill

1881 Smoke Abatement Exhibition opens in South Kensington, then travels to Manchester the following year

1882 Fog and Smoke Committee renamed as the Smoke Abatement Institute, then as the National Smoke Abatement Institution

1882 Metropolitan Public Gardens Association established in London by Lord Reginald Brabazon (later the Twelfth Earl of Meath)

1896   Swedish scientist Svante Arrhenius suggests that coal
        combustion might cause global warming

1898   Coal Smoke Abatement Society (CSAS) established in
        London by Sir William Blake Richmond, RA

1909   Smoke Abatement League of Great Britain (SALGB) established,
        with branches in Glasgow, Manchester, and Sheffield

1921   Miners' strike and resulting coal shortage brings temporary
        improvement to air quality

1927   Atmospheric Pollution Research Committee established by
        UK government

1929   National Smoke Abatement Society formed through a merger
        of the CSAS and the SALGB

1951   Coventry establishes first smokeless zone in Britain

1952   Intense smog kills thousands in London

1953   British government appoints Beaver Committee to investigate
        smog disaster

1956   Clean Air Act enacted, and government orders hundreds of
        localities to establish smoke control areas

1965   Natural gas discovered in North Sea, leading to the end of
        coal gasification in Britain

1990   Intergovernmental Panel on Climate Change announces that
        fossil fuel emissions are causing global warming

2003   London establishes Congestion Charge Zone, designed to
        reduce traffic and vehicle emissions

2015   Kellingley colliery, the final deep coal mine in Britain, closes

2015   Paris Climate Accord reached, through which 195 countries
        agree to reduce their carbon emissions

Chapter 1

Coal, Smoke, and History

It will be a task for the future social historian to explain why the English of
our time were content to live in dirty and gloomy air.
—*John W. Graham, 1907*[1]

To millions of our town-dwellers smoke is just what comes out of the chimney,
as coal is just what goes on the fire. The idea that smoke is a "problem,"
something to be prevented, simply does not exist.
—*Arnold Marsh, 1947*[2]

Around the world, a growing number of people are asking questions about
how technology is affecting the natural world, human health, and society.
Fierce debates rage over whether current levels of consumption and pollution are sustainable, and whether it is possible to both protect the environment and create material prosperity. These vital questions, which will
become even more pressing in the decades ahead, have a forgotten history.
Of all the challenges that confront the world today, few threaten as many
people as the pollution that results from burning fossil fuels. Three billion
people—half of the world's population—now live in cities, many of which
contain air that is unfit to breathe. Two hundred years ago, however, only
one city on the planet used significant quantities of fossil fuels and experienced the pollution that such consumption entails. In 1800 Londoners
burned one million tons of coal—an amount that was equivalent to a ton
for every resident. From that year forward, fossil fuel consumption skyrocketed throughout the country, literally fueling Britain's rise as the most
powerful manufacturing, trading, and imperial power that the world had
ever seen.

Many substances that are now viewed as serious pollutants, including asbestos, lead, and CFCs (chlorofluorocarbons), were once considered innocuous. The same was long true of the products of coal combustion. By the middle of the nineteenth century coal smoke filled many British cities, yet few people saw it as detrimental to either human health or to the wider environment. In their view, pollution came not from energy use or industry, but from natural biological processes. They blamed disease on miasma, an invisible gas thought to be given off by decaying plant and animal matter. Thus, the most polluted environments were those in which the greatest quantities of decomposing biomass were found: marshes, jungles, graveyards, cesspools, and sewers. Many people not only considered coal smoke to be harmless, but actually thought of it as an antidote to pollution. According to miasma theory, the acids and carbon in smoke were powerful disinfectants.

The notion that coal smoke was beneficial to health began to change during the late nineteenth century. As the air of British cities and towns filled with ever-denser smoke, scientists coined new terms such as acid rain and smog, and physicians blamed smoke for a range of health impairments, including respiratory diseases, rickets, decreased stamina, and even "racial degeneration." At the same time that these changes were occurring, the new science of bacteriology was leading many to abandon the belief that disease came from miasma. The conceptual disappearance of miasma not only changed attitudes and policies toward public health; it also removed a major justification for coal smoke.

Britain, the "first industrial nation"[3] and the first to become predominantly urban, was also the place in which the modern idea of pollution was invented. During the nineteenth and first half of the twentieth centuries, people in Britain came to understand coal smoke as pollution and came to understand pollution as an entity that should be regulated by the state, a state that would eventually—through the Clean Air Act of 1956— reach into people's homes and extinguish the coal fires that had warmed their hearths for generations. This book tells that story.

People in Britain began using coal well before the industrial revolution. In contrast to the situation in many other parts of the world, where coal can be found only deep underground, in Britain substantial quantities of coal lay near the surface. Nearly two thousand years ago, during the Roman occupation of Britain, people dug shallow pits to remove the mineral from the ground, but the rudimentary nature of early mining techniques long restricted the amount that could be extracted. Another limiting factor was transportation. In an age before canals and railroads, the difficulty and cost of transporting coal over land was enormous. Starting in the thirteenth century, however, coal began to be mined near the port city of Newcastle-upon-Tyne in northeastern England. From there it could be transported via ship to other coastal cities. By far the largest market, for coal and all other commodities in Britain, was London. Because of how coal from Newcastle reached them, Londoners long referred to it as sea-coal.

Initially, much of the coal burned in London was used to heat kilns that converted limestone into lime, an essential ingredient in mortar. Over time, other industrial activities, such as metal smithing, also began to use coal. Wood continued to dominate the market for household fuel, not only because the smoke it produced was considered more pleasant than that from coal, but also because wood could be brought indoors without scattering dust and grime. Between 1540 and 1640, the price of firewood in London nearly tripled relative to the cost of other goods, while the price of coal rose no faster than the overall rate of inflation. By the middle of the seventeenth century, coal was the dominant fuel in London for domestic as well as industrial uses.[4]

The English inventor Thomas Newcomen (1663–1729) produced a working steam engine at the beginning of the eighteenth century. Although it was incredibly inefficient and produced little power, the Newcomen engine revolutionized coal mining by making deep mines possible. For in addition to providing mechanical power to hoist heavy loads of coal to the surface, steam engines drove pumps that prevented mines from becoming flooded with seeping groundwater. The chemical energy stored in coal, when transformed into heat and mechanical energy by the steam engine, made it possible for miners in Britain to reach and extract a seemingly endless supply of coal from the earth.

During the last quarter of the eighteenth century the Scottish inventor James Watt (1736–1819) greatly improved the efficiency and power of the steam engine.[5] Watt's innovations, and those of other engineers who made further improvements, made steam power an attractive source of power in industry and thus boosted demand for coal. By the early nineteenth century steam engines were sufficiently compact and powerful to become mobile; the steamships and railroad locomotives that followed not only consumed enormous quantities of coal, but also allowed coal to be transported to places that previously had been forced to rely on renewable sources of energy.

Neither the presence of coal nor the existence of steam engines was alone sufficient to make Britain the wealthiest and most polluted country in the world during the nineteenth century; together, they changed everything. As mill owners adopted steam in place of animal and water power, demand for coal rose sharply. In 1800 approximately 15 million tons of coal were burned in Britain. Coal use increased dramatically during the nineteenth century and continued rising until the eve of the First World War, when Britain's coal consumption reached an all-time high of 183 million tons.[6]

At the beginning of the twentieth century Britain remained the largest producer of coal in Europe, with an annual output of 229 million tons. As its neighbors industrialized, their coal production rose rapidly. The German lands, which had extracted just 6 million tons of coal in 1850, produced 43 million tons in 1871, the year of German unification. In 1900 Germany produced 150 million tons, followed by France (33 million tons) and Belgium (24 million tons). Although British fears of industrial decline—or at least loss of predominance in coal—focused on Germany, the greatest challenge came from across the Atlantic. The United States, which had mined just 8 million tons of coal in 1850, produced 245 million tons in 1900, making it the world's leading coal country.[7]

The adoption of coal-burning steam engines not only made it possible for factories to increase production; it also freed them from the geographical and seasonal constraints inherent in the use of water power. Instead of remaining dispersed across the countryside, factories became concentrated near coal mines and coal transportation routes, creating urban centers where workers and consumers lived. At the start of the First World War over one million people worked in coal mines in Britain, and the work of transporting, distributing, and loading it into boilers, furnaces, fireplaces, and kitchen ranges involved millions more.[8] People relied on coal to fuel industry, to power railroads and ships, to keep warm, and to cook. Coal was also used to make gas, which was the primary source of indoor and street lighting in the nineteenth century. Some, including the English economist and logician William Stanley Jevons (1835–82)—famous for his 1865 prediction that Britain would run out of coal within decades—claimed that the nation's industrial and imperial ascendancy came not so much from hard work and sound government as from its coal.[9]

The burning of coal released not only energy, but also large quantities of smoke, soot, and acidic vapors. Bituminous coal, the most prevalent form of coal found in Britain, contains a large proportion of impurities. On average, 20 percent of its weight consists of sulphur, volatile hydrocarbons, and other chemicals. Even under ideal conditions the burning of bituminous coal produces toxic ash, sulphur dioxide (a key ingredient in acid rain), and the greenhouse gas carbon dioxide. Optimal combustion requires a high temperature and a precise ratio of oxygen to fuel. These conditions rarely existed in practice.[10]

Much of Britain's coal consumption, and the smoke that accompanied it, was concentrated in urban areas. Industrialization caused the air of large towns and cities—already heavily polluted by household coal use—to deteriorate much further. By 1851 more people in Britain lived in towns and cities than in the countryside—something unprecedented in world history. As a result of this urbanization, many cities experienced population growth that was substantially higher than the national average. Glasgow, Leeds, and Sheffield, for example, each grew nearly tenfold during the nineteenth century.[11] The growth of London was proportionately smaller, but in numerical terms its expansion vastly exceeded that of all other cities in Britain, increasing from approximately one million inhabitants in 1800 to over six million a century later.[12] Describing London in the 1830s, one writer was struck by the "dense canopy of smoke that spread itself over her countless streets and squares, enveloping a million and a half human beings in murky vapour."[13] In 1913 Londoners burned over 15 million tons of coal—an average of two tons for every man, woman, and child. Other cities also burned mountains of coal. Manchester, with less than a tenth of the population of London at the start of the twentieth century, used nearly half as much coal per year.[14]

Sporadic complaints about coal smoke began soon after coal first arrived in London; in the 1280s two royal commissions were appointed to investigate the matter. As a result of cheaper firewood becoming available—and people becoming accustomed to coal smoke—complaints about smoke soon subsided and remained at a low level until the upsurge in coal use that occurred in the late sixteenth and early seventeenth centuries.[15] In 1661 the diarist John Evelyn (1620–1706) published *Fumifugium, or the Inconvenience of the Aer and Smoake of London Dissipated,* a strongly worded tract that remains one the most famous denunciations of air pollution ever written.[16] Despite Evelyn's forceful attack on coal smoke, his position remained an isolated one for the following two hundred years.

Although numerous cities in Britain were extremely smoky by 1800, few people at that time viewed coal smoke as a problem, and no one used the word pollution to describe it.[17]

During the second half of the nineteenth century, however, public health experts, urban reformers, and journalists redefined coal smoke, transforming it from an accepted part of the urban environment into a problem. The redefinition of smoke as pollution was both a scientific and a social process. At the same time that researchers were analyzing the particles and vapors that issued from smokestacks and chimneys, others were talking about the subject in meeting rooms, newspapers, and magazines. What caused this transformation? According to some, the amount of smoke in the air had simply reached an intolerable level. This claim served an important rhetorical purpose for smoke abatement activists who sought to place it at the center of problems that ought to be corrected, but it fails to explain why smoke was considered undesirable. The reimagining of pollution reflected changes not only in the natural environment, but also in scientific understanding, political ideology, and popular culture.

Many people in Britain believed that industrialization and urbanization were causing serious social, health, environmental, economic, and strategic problems. Technology had re-created the world through developments such as railroads, massive bridges, and transoceanic telegraph cables, but it also provoked great anxiety. A contradiction seemed to exist between the capacity of humanity to transcend formerly insurmountable environmental limitations and people's inability to anticipate or control the consequences of their new technologies. By 1900 many viewed coal smoke as the embodiment of these concerns. Britain was increasingly unable to grow enough food to feed its people, remote rural places were becoming coated with urban grime, the amount of sunlight was declining, particular varieties of animal life were disappearing, and trees were becoming stunted by a nightmarish new substance: acid rain.[18] The latter was discovered in the 1850s by the chemist Robert Angus Smith (1817–84), who spent the final twenty years of his life as the director of the Alkali Inspectorate, the world's first national pollution control agency. Although this agency initially had responsibility for only a single industry (and a single chemical), its purview eventually expanded to cover many other trades and substances. For roughly a century, the Alkali Inspectorate functioned as Britain's primary environmental regulatory agency.[19]

Concerns about environmental degradation were also connected to anxieties about cultural decline.[20] Idealizing Britain's medieval past, some urged a return to a simpler form of society, closer to nature and free from smoke. Inverting the notion that urbanization and industrial development were proof of progress, radical artists and writers such as John Ruskin and William Morris asserted that England was sacrificing its connections to nature and the past in a misguided quest for material gain.[21] They maintained that the whole country was becoming subordinated to the demands of cities and industry, and that rural areas were losing their "natural" character. Others argued that environmental degradation was causing the people of Britain—particularly the urban poor—to degenerate both physically and morally. According to the proponents of this view, inadequate sunlight and fresh air would lead not only to individual degeneration, but also to national decline. Strong, healthy, and hardworking citizens, they argued, were essential to Britain's industrial and imperial power.[22] Another group maintained that coal smoke was a dangerous contributor to social and political unrest. Air thick with smoke seemed to provide an ideal cloak for crime, immorality, and mob activity. And because prevailing winds concentrated smoke in particular districts, smoke accentuated the residential segregation of different classes. In addition to making it difficult for members of the middle and upper classes to monitor what the "masses" were up to, it also created resentment among poor people who lived in polluted areas that they could not afford to leave.[23]

Smoke provided reformers who had divergent goals with a unifying symbol of the need for change. These concerns led to vocal antismoke activism in many British cities, most prominently by the Coal Smoke Abatement Society, which focused on London, and by the Smoke Abatement League of Great Britain, which sought to clear the air in industrial districts. During the late nineteenth century many of these reformers concluded that something more than education and gentle persuasion was needed to rid the air of smoke. Shaped in part by the ideology of the "new liberalism," which envisioned a much more active role of the state in economic matters, a growing number of reformers advocated the passage of comprehensive national legislation on smoke, coupled with vigorous (and perhaps national) enforcement. Some went so far as to call for restrictions on smoke from private houses, which had always been exempted from regulation. Despite the efforts of some in Parliament, the majority of MPs long resisted these arguments.

Reformers achieved greater success, however, in getting the national government to recognize air pollution as an important subject of research. During the First World War the Meteorological Office began funding the

Committee for the Investigation of Atmospheric Pollution, an independent group established in 1912 to coordinate the measurement of soot deposits in many of the smokiest cities in Britain. In 1927 the committee was absorbed by the Department of Scientific and Industrial Research and renamed the Atmospheric Pollution Research Committee. Although soot deposits provided only a rough approximation of air quality, they made it possible to compare different places and to track changes over time.[24]

The solutions that smoke abatement activists proposed depended on the ways in which they conceived of pollution. Those who disliked coal smoke because of an antipathy to modernity or industrialism were apt to focus their attention on smoke that came from factory smokestacks rather than household chimneys. Despite the prominence of industrial smoke, however, a large proportion of the smoke that filled the air of Victorian Britain came from the familiar domestic fireplace. As a result of their ideological predisposition against industry, many activists failed to recognize that smoke would not be banished by either rejecting or reforming the industrial system, but would require people like themselves to modify the ways that they used coal. Industrialists and their supporters, on the other hand, often agreed that smoke was a problem but denied causing it. They pointed out that factories, in contrast to private houses, already faced legal restrictions on smoke emissions, and they argued that further mandatory reductions in industrial smoke were unfair and impossible to achieve. When confronted about the smoke that their factories produced, owners often shifted the blame to their employees.

Many historians have written about the ways in which industrialization and urbanization transformed the economic and social life of Britain, but far fewer have examined how they affected the environment, reshaped ideas about nature, or stimulated the rise of environmental activism.[25] Yet the histories of country and city, of natural and built environments, are in fact deeply interwoven. Cities are concentrations not only of people and production, but also of consumption. Without outside supplies of air, water, food, and energy, urban life would be unsustainable.[26]

In contrast to the large and impressive body of scholarship on the history of air pollution in the United States by Joel A. Tarr, Martin V. Melosi, and numerous others, relatively few books have been published about Britain's experience with air pollution.[27] Eric Ashby and Mary Anderson's *The Politics of Clean Air* (1981) focuses on Parliament and has relatively little to say about either cultural attitudes or municipal enforcement of the law. Peter Brimblecombe's *The Big Smoke: A History of Air Pollution in London since Medieval Times* (1987) provides an important analysis of changes in the quantity and concentration of coal smoke over time, but it does not examine how the meaning of smoke changed over time. Stephen Mosley's *The Chimney of the World: A History of Smoke Pollution in Victorian and Edwardian Manchester* (2001), although an excellent study of both popular and elite understandings of coal smoke in that city, does not address what happened in London or the rest of Britain, and it ends in the early twentieth century.[28]

In his classic essay, "Ideas of Nature," Raymond Williams observed that the ways in which people think about the environment reveal a great deal about how they interact with each other and with the natural world.[29] This book examines not only the tangible reality of pollution, but also its place in people's minds. In attempting to understand the history of air pollution, it is not enough to know what substances, in what quantities, entered the environment in the past. Equally important are the attitudes, ideologies, and perceptions that led to the creation of these pollutants and that structured people's understanding of their effects.

Chapter 2

The Miasma Era

Go into the country by all means for fresh air; but do
not take for granted that you will get it.
—**The Builder,** *1859*[1]

IN 1800 MOST PEOPLE in Britain blamed impure air not on technological processes, but natural ones. They believed that the most serious con-
taminant in the environment was *miasma,* an airborne substance thought
to be produced by decomposing biological material. Miasma was considered dangerous because it was thought capable of causing the bodies of
people who inhaled it to decompose or ferment. When this happened, individuals were said to suffer from a zymotic illness, from the Greek word
for fermentation.[2] According to some, even a single breath of impure air
might be enough to cause illness and death. Sources of miasma appeared
ubiquitous; contemporaries traced it to stagnant marshes, garbage, horse
manure, burial grounds, and the products of human respiration, perspi-
ration, and excretion.[3]

Miasma theory suggested that the natural environment abounded with
hazards. As the physician Thomas Southwood Smith (1788–1861) put it in
1830, "Nature, with her burning sun, her stilled and pent-up wind, her
stagnant and teeming marsh, manufactures plague on a large and fearful

Figure 2.1. "Sweetening the Air of the Country." This depiction of a village in Essex accompanied an article that warned people to avoid polluting the air near their homes with miasma from decomposing household wastes. From *Builder*, 23 July 1859, 488.

scale."[4] The more biomass a particular place contained and the greater the temperature, the more miasma was thought to be produced. For this reason the tropics—characterized by one British observer in 1862 as places "where vast quantities of organic matter, the *débris* of a luxuriant vegetation, are rapidly passing into decomposition"—appeared to be particularly dangerous.[5] In apparent confirmation of this theory, European colonists, soldiers, and missionaries experienced appalling mortality rates in sub-Saharan Africa, which was often dubbed the "white man's grave."[6]

Many attributed the dangers of tropical regions not only to their climate, but also to the failure of those who lived there to exert proper control over nature. Miasma theory thus provided a convenient justification for colonialism. According to this logic, only Europeans possessed the medical knowledge and the organizational ability necessary to domesticate the jungle. James Lane Notter (1843–1923), an instructor at the Army Medical School, warned in 1880 that vegetation would exert a beneficial effect on health only if it were "carefully attended to and kept within certain limits. In any climate and under any circumstances the exuberant growth of plants and trees is bad, especially in the tropics, where rank vegetation abounds." Such "rapid and uncared-for growth," he insisted, led directly to "decay taking place, with all its attendant products of decomposition, poisoning the air, and rendering, by its noxious vapours and mists, the atmosphere unendurable."[7] In the view of such experts, the essential difference between healthy and unhealthy nature—as between an ordinary worker and a member of "the dangerous classes"—was the extent to which each was supervised and disciplined. Victorian sanitarians often maintained that their expertise was vital to prevent unhealthy environments from poisoning the air with miasma. Proponents of imperialism often used medical arguments as justification to exert British control not just over "unhealthy" regions, but also their inhabitants. Civilization, in their view, would domesticate not only unruly and dangerous nature, but unruly and dangerous natives.[8]

Although miasma theory raised the disquieting prospect that impure air might prove fatal, it also suggested that people could use their own senses to protect themselves from unhealthy conditions. As Edwin Chadwick (1800–1890), Britain's leading public health expert, put it in 1846, "All smell is disease."[9] Later observers have expressed surprise at the apparent complacency with which Victorians dumped untreated sewage into rivers that were also used as a source of drinking water. Chadwick and his contemporaries, however, believed that sewage posed a much greater risk to air than to water.[10] According to this logic, dumping sewage into rivers was preferable to allowing it to accumulate near habitations, where it would stagnate and contaminate the air with miasma. When people did express concern about the presence of sewage in watercourses, they typically identified the problem as one of impure air, rather than impure water. Commenting on the "great stink" of 1858, when warm weather and an absence of breezes intensified the pungency of the sewage-laden Thames, the Society for Promoting Christian Knowledge noted that the episode had "polluted the air."[11]

Living Things

Living organisms were also thought to be a source of polluted air. The *Times* claimed in 1855 that "the vitiation of the air by domestic animals kept in the house is very considerable. The keeping of such animals in small houses ought not to be tolerated. Birds consume a very large quantity of oxygen, and the excrements of these and of domestic animals generally increase the poisonous effect of their presence."[12] Even houseplants were thought to pose a danger. Asserting that living vegetation expelled carbon dioxide at night, an 1859 pamphlet warned, "Those pretty plants

Figure 2.2. Many people in nineteenth-century Britain feared that sewer gas contained disease-causing miasma. In this illustration, miasma is shown to pose a deadly danger not only to slum dwellers, but also to the middle class. From *Punch*, 1 Mar. 1851, 83.

. . . which you have put in the bed-room window, to look cheerful and bright, rob you of good air when you are asleep, for they have a way of breathing and want air just as we do; they will be much better down in the kitchen."[13] Concerns about the potentially harmful nighttime exhalations of plants were not limited to indoor spaces. While some believed that vapors released by vegetation made outdoor air harmful after dark,[14] most medical experts rejected this view. As one physician put it in the 1870s, "The night air in towns is often the purest. In some country places, as in the Fens, there might be malaria, and agues [fevers] might be the result of exposure to it, but in most towns . . . there is no danger in breathing night air from without; the real danger is from night air within doors."[15]

Commentators expressed even greater concern about the effect of the human body on the air. As the president of the British Medical Association noted in 1879, "The higher the animal life is in the scale of creation, the more injurious is the excreta of such animals, and that of man most of all."[16] Although opinions differed as to whether the exhalations of healthy individuals could induce disease, most believed that a person suffering from a zymotic illness posed a clear risk to others. Such individuals were considered dangerous because they were thought to be undergoing internal fermentation or decomposition. According to Thomas Southwood Smith, a poorly ventilated room containing a fever patient was "perfectly analogous to a stagnant pool in Ethiopia, full of the bodies of dead locusts. The poison in both cases is the same; the difference is merely in the degree of its potency."[17] Although miasma could not be seen, many people believed that its presence was often betrayed by fog.

Fog

Fog exists at the intersection of nature and culture. Because it floats freely between country and city, it has long reminded people of the connections between the two. Fog was often seen as a mysterious substance that appeared and departed with little warning, and that dramatically altered the appearance of reality. An 1853 article in the *Leisure Hour* embraced the ability of fog to transform London into a strange and mythical landscape, with

> visions, looming rapidly into view, and as rapidly disappearing—of monstrous moving mountains, drawn by mammoths and megatheriums, and driven each one by a shadowy colossus of Rhodes, and crowned with other colossi, sprawling in attitudes absurdly familiar, considering their immensities, on the top. These we know well enough to be the omnibuses, magnified in the gloom. Every man we meet, indeed, is magnified to ten times his proper size at a distance, and only dwindles down to human dimensions as he rubs shoulders with us, and is gone.[18]

Figure 2.3. "Nobody that knows them could doubt the respectability of these two gentlemen, yet you would hardly credit the unnecessary panic their imaginations caused them the other night in the fog!" Fog caused familiar places to appear strange and made it difficult for people to distinguish friend from foe. Cartoon by Linley Sambourne in *Punch,* 19 Feb. 1870, 72.

Although some observers welcomed fog, most considered it dangerous. Well before Sir Arthur Conan Doyle used fog as an essential dramatic element in his Sherlock Holmes novels, observers noted that its opaque nature provided criminals with a smoke screen. As a popular periodical noted in 1855, "A London fog is a very carnival of petty larceny." Pickpockets, cloaked in the fog, "will contrive to relieve you of any loose cash, pocket-book, or tempting 'ticker,' with a dexterity you cannot but admire, however much you may rue your loss."[19] Many also blamed fog for causing accidents. For lack of visibility ships ran aground, pedestrians were struck down by moving vehicles, carriages collided, and people drowned by falling into canals and harbors. During a week of thick fog in December 1891, eight people died in London after stumbling into the water, none of whom were suspected of intoxication or suicide.[20]

The most serious danger from fog, in the opinion of most people, was neither crime nor accidents, but disease. Fog was thought to come from the same kinds of places that produced miasma. Swamps and marshes, partly because they were not subject to human management, were often blamed for poisoning the air with "unwholesome" substances.[21] Another reason that such places were feared was malaria, which had long been prevalent in wetland areas, even as far north as England.[22] The belief that this disease came from breathing impure air was embodied in its name, which means bad air in Italian.[23] An 1859 book ridiculed as superstitious the notion that malaria came from mosquitoes. Its author, an English physician, recalled that on a visit to Rome—a city notorious for its malaria— he had been reading one night in front of an open window. A frightened servant pointed out that his lamp was attracting insects into the room and gravely warned him that they might cause malaria. The physician smugly dismissed these fears as baseless, asserting that malaria was not spread by insects but by "the condition of the air."[24] Although some mid-nineteenth-century scientists suspected that malaria was transmitted by mosquitoes, conclusive evidence of this was not discovered until 1897.[25]

For most of the nineteenth century, fog was feared as a threat to health because of its association with malodorous swamps. Most believed that fog was incapable of originating in urban areas, but instead formed in barren and inhospitable "wastelands." Reflecting this, the appearance of fog in cities was typically described as a "visitation." As one writer declared in 1853, "Science informs us that the cause of fog is the defective drainage of the lands and marshes, extending for miles on the banks of the Thames, south and east of the city."[26] Two years later the *Times* warned that "a breath of marsh vapour may poison a whole neighbourhood for weeks."[27] The medical professor Edward Greenhow (1814–88), who supported this view, thought it significant that a major influenza epidemic in London had been "preceded by mist and gloominess."[28] As another writer observed, the appearance of fog served to warn people of the presence of invisible miasmatic gases, "arising from the same cause . . . , which are always more or less prevalent."[29] Fog was miasma made visible.

Smoke

Because they assumed that impure air came from biological sources, most people—until the late nineteenth century—viewed coal smoke as benign. In the words of the historian Alain Corbin, "What was intolerable was the odor of putrefaction or fermentation, not of combustion."[30] Believing that aerial contact with decomposing substances could cause disease by making the body decompose as well, they developed strategies designed to ameliorate "miasmatic vapours." Coal smoke was an important ally in this struggle. For centuries, smoke was used not only to preserve meat, but also to prevent disease by acting as a fumigant against airborne impurities. The carbon and sulphur it contained were seen as fumigants that could neutralize miasma. Within this context, people looked at coal smoke not as pollution, but as something that could help to protect against it.[31] When plague struck London in the sixteenth and seventeenth centuries officials urged residents to build bonfires of coal to disinfect the air.[32] Not everyone agreed with this advice, however. Writing in the 1660s, John Evelyn rejected the argument that coal smoke acted as a beneficial fumigant. Instead, he claimed, it contaminated the air with "an impure and thick mist, accompanied with a fuliginous and filthy vapor."[33] Despite critics such as Evelyn, most people believed that smoke, no matter how unpleasant it might be, caused no harm to health.

The notion that smoke could prevent or even cure disease remained widespread in the nineteenth century. The surgeon John Atkinson suggested in 1848 that tuberculosis sufferers should inhale coal smoke and other chemical gases. In his view, creosote, pitch, tar, and naphtha would arrest the progress of the disease.[34] Even critics of smoke sometimes expressed concerns that reducing it might increase the prevalence of infectious disease. In an 1859 article in which he coined the term acid rain, the chemist Robert Angus Smith noted that although the sulphurous acid in coal smoke damaged buildings, it also acted as a disinfectant against "putrid matter in towns." Smith, who had studied under the eminent German chemist Justus von Liebig and worked closely with the public health authority Edwin Chadwick, added that British cities generally smelled better than those in other countries because Britain's air contained more sulphur.[35] The lord mayor of London, an honored speaker at the opening of an antismoke exhibition in 1881, went so far as to suggest that smoke abatement might harm public health. He claimed that although malarial fever had once been common "in the marshes of the Thames, . . . after the erection of the great factory chimneys the disease had not affected people living in that neighbourhood."[36]

Industry had a strong incentive to perpetuate the view that airborne chemicals, including carbon and acidic vapors, were beneficial. In the 1850s the Manchester manufacturer Peter Spence maintained that the carbon in coal smoke was "guiltless of any deleterious effect on human health, is one of the most anti-putrescent bodies, and while floating in the atmosphere, does all it can to arrest and destroy noxious and miasmatic vapours."[37] This valuable property of smoke could even be used to disinfect sewage, he argued, "by introducing the products of combustion from chimneys into the sewers."[38] The trade journal *Chemical News,* which did much to publicize Spence's views, claimed in 1862 that even though coal combustion caused acid rain, this substance was "rather beneficial to health than otherwise, as tending to retard the putrefaction of animal matter on which it falls."[39]

When people did advocate smoke prevention prior to the late nineteenth century, they typically emphasized economic reasons for doing so, not health or environmental considerations. According to many observers, the most troubling aspect of smoke was the waste that it represented. Inefficiency was a key concern of a trade group formed in 1854, the Manchester Steam Users' Association for the Prevention of Steam Boiler Explosions and for the Attainment of Economy in the Application of Steam.[40] Using language that foreshadowed that of later conservationists, the physician and inventor Neil Arnott (1788–1874) argued in 1855 that the inefficient use of coal was not simply a concern for individual householders and factory owners: "Coal is a part of our national wealth, of which, whatever is once used can never, like corn or any produce of industry, be renewed or replaced." Wasteful consumption of coal, he insisted, "is not a slight improvidence but a serious crime committed against future generations."[41]

The view that smoke was primarily an economic concern long remained popular, even among medical doctors. In an address on coal to a public health group in 1872, the speaker confined his remarks almost entirely to questions of economics and convenience and said almost nothing about the effect of coal smoke on health.[42] Although some experts in the middle years of the nineteenth century suggested that smoke caused health problems, they lacked compelling evidence to support this hypothesis. The medical officer John Simon (1816–1904), a prominent critic of smoke, noted that it was "at the head of the class of evils . . . which are complained of on the ground of their offensiveness rather than of their injury to health—as nuisances rather than poisons."[43]

CHAPTER 3

Pollution Redefined

The purest air is to be found on mountains, moors, or far away from
contaminating and polluting agencies, such as aggregations of men
and animals, manufactories, etc.
—*Cornelius B. Fox, 1878*[1]

DURING THE FINAL DECADES of the nineteenth century a new understanding of air pollution began to develop in Britain. Instead of seeing nature as dirty and civilization as a source of cleanliness, many people began
to assume that nature was inherently pure and only became unhealthy as
a result of technological processes. Important as this change was, significant continuities connect the two views. First, any deleterious effects from
both miasma and coal smoke were blamed on the incomplete oxidation
of matter. Under ideal conditions, most people believed, neither biological decay nor combustion would pose any dangers. All that would remain
in either case would be simple, inert substances. Second, both miasma and
smoke were associated with fog. During the early nineteenth century fog
was thought incapable of forming in cities, and its presence in them was
seen as a case of the country polluting the city with poisonous miasma
from rotting swamps. Over time, however, many people began to associate
fog not with marshes and miasma, but with cities and coal smoke. Urban
residents came to conflate fog with coal smoke, for the very conditions

that made fog likely—cool temperatures and still air—increased the amount of coal that people burned to keep warm and prevented the resulting smoke from dispersing.[2]

Germs

Miasma theory faced a strong challenge in the late nineteenth century from the germ theory of disease. Proponents of the germ theory attacked sanitarians as not only misguided, but dangerous in that they created a false sense of security. Bacteriologists argued that because disease germs were invisible and odorless, surfaces that looked clean or air that smelled fresh might nonetheless harbor deadly microbes. They asserted that all disease could be explained in terms of germs, and that only they possessed the expertise needed to determine whether germs were present.

Looking back at the remarkable discoveries in bacteriology that occurred in the late nineteenth century, historians have long disagreed about their significance to public health. While some credit the germ theory for leading medical experts to focus on germs, the true source of disease, rather than chase red herrings such as miasma, others, particularly social historians, disagree. They argue that the discovery of bacteria offered few tangible benefits to patients until the advent of antibiotics decades later, and they suggest that the efforts of sanitarians, even if based upon erroneous ideas, nonetheless reduced the incidence of disease in the decades that preceded the advent of germ theory. Regardless of their position on the impact of bacteriology, both sets of historians share the view that it led people in late nineteenth-century Britain to stop paying attention to environmental conditions.[3] The problem with this interpretation is that it focuses overly much on what physicians were doing and saying, while ignoring the ways in which sanitarians appropriated the germ theory in their ongoing struggle for relevance. Rather than abolishing the long-standing dividing lines and disputes in the debate over the role that the environment played in health and disease, the germ theory instead cast it in a new light.

Most sanitary reformers vehemently rejected the charge that they had nothing to contribute to Britain's health. While a few simply ignored the germ theory in the hope that it was a passing fad, others responded by recasting pollution as something that came not only from plants and animals, but from technology. Instead of focusing on decaying organic matter and urging the building of sewers, they shifted their attention to coal smoke and called for its abolition. The identification of smoke as a threat to health gave public health reformers a renewed sense of mission at a time when two forces—the growing professionalization of medicine and the germ theory—were forcing them to cede primary responsibility for the prevention of infectious diseases to physicians. Through their efforts, an environmental interpretation of disease persisted, albeit in a form very different than in the past. Fog, once feared because of the miasma it was thought to carry, now became a concern because of its association with coal smoke. Overturning the long-standing assumption that smoke acted as a valuable disinfectant, many public health activists decried it as a dangerous source of impure air. Some, such as the meteorologist and sanitary reformer William Napier Shaw (1854–1945), went so far as to describe coal smoke as aerial sewage.[4]

This mixture of biological and technological metaphors was intended to convince people that smoke-filled air was no less serious a threat to health than polluted water, yet it could also lead to confusion. For although impure water had been a major concern in the middle of the nineteenth century, most people believed that the problem had been solved—thanks to the construction of an integrated network of sewers. Language intended to stress continuities with an earlier problem might cause some to miss the point. Joe St. Loe Strachey, the editor of the *Spectator*, articulated this concern in 1898. Writing to the indefatigable reformer Thomas Coglan Horsfall (1841–1932) in the same month that a new environmental pressure group, the Coal Smoke Abatement Society, formed, Strachey wrote, "I am very keen about preserving the amenities of England i.e. by fighting smoke and the pressing wk [?] of open spaces. . . . Do write me a modest letter to the editor . . . pointing out that the race is and will degenerate unless we maintain better conditions. Don't use the word '*sanitary*' too much or indeed if possible at all, as the idiot publisher will think it means drains and cesspools and W.C.s and nothing else."[5]

Although most sanitarians acknowledged that germs had the potential to cause disease, they insisted that environmental conditions affected both their virulence and people's ability to resist them. While serving as president of the British Medical Association in 1879, Alfred Carpenter (1825–92) asserted that exposure to germs was not sufficient to cause illness. Equally important, he argued, were environmental conditions that could make the body more susceptible to their action.[6] The president of the Sanitary Institute, a professional association for public health inspectors, also rejected the claim that the discovery of germs made environmental reform irrelevant to health. Speaking in 1898, Joseph Fayrer noted,

"It is not enough that we know the seed, but it is necessary that we should also know the nature of the soil, the meteorological and other conditions which determine whether it is to grow and multiply or to remain inert and harmless."[7]

Not all scientists were persuaded that smoke was a form of pollution. Addressing the Royal Society of Edinburgh in 1880, the Scottish physicist John Aitken "drew attention to the deodorising and antiseptic powers of smoke and sulphur, which probably operated beneficially in killing the deadly germs, and disinfecting the foul smells, which cling about the stagnant air of fogs, and suggested caution lest, by suppressing smoke, we substitute a greater evil for a lesser one."[8] In the years that followed, a growing number of scientists and physicians criticized Aitken's position. The same year that Aitken spoke, the *Lancet* argued that the high prevalence of infectious diseases in London raised serious doubts about the hypothesis that smoke destroyed germs.[9] F. S. B. François de Chaumont, a prominent medical professor, declared in 1885 that an atmosphere containing enough sulphurous acid to kill bacteria would be fatal for human beings.[10] These new ideas reached a wide circulation. In 1892 a popular magazine asserted that few people continued to believe that coal smoke helped to prevent disease:

> Formerly, there were always persons to be found who . . . were able to inhale the disinfecting particles with which a London fog is charged, and to feel themselves the better for the experience. We have heard no such tales of the last fogs. No disinfecting action can be traced to them. A strong chest constituted no protection against their assaults. The air had somehow ceased to be fit for human consumption. It produced not coughs merely, but headaches, nausea, and other ordinary accompaniments of small doses of poison.[11]

Of even greater concern than smoke's harmful effects on the respiratory tract and its apparent inability to destroy germs was a growing suspicion that smoke interfered with what many viewed as "nature's disinfectants."[12]

Ozone

Fresh air has long been viewed as a disinfectant. Following the discovery of oxygen in the late eighteenth century, many credited it with the ability to purify the air by burning up the miasma it contained. According to this view, poorly ventilated rooms were unhealthy because they contained too little oxygen to destroy miasma.[13] These ideas were reinforced in the middle of the nineteenth century when the German scientist Christian Friedrich Schönbein (1799–1868) discovered ozone.[14] Ozone is a form of oxygen that consists of three atoms instead of the usual two. This arrangement makes ozone unstable; it readily oxidizes any flammable molecule it encounters. Because of its reactivity ozone was assumed to be an ideal substance with which to rid the air of miasma. As one writer observed in 1865, ozone "plays a very important part in the economy of nature" by purifying urban air containing "organic matter in a high state of decomposition or putrescency."[15]

Measurements of atmospheric ozone began in the early 1850s. These observations showed that ozone was plentiful in the countryside and could always be found in air blowing in from the sea; conversely, the air of large cities appeared to contain little of the molecule.[16] The presence of ozone in air far from cities led many to conclude that industrial society was at odds with an inherently healthy nature. In a lecture sponsored by the Manchester and Salford Sanitary Association in the 1870s, the chemist John Angell used romantic language to elaborate: "By the side of the waterfall, on the vast moorland plain, by the laving of the sea-wave and in the track of the lightning flash . . . , [the] health-establishing powers [of oxygen] are greatly intensified, and we then know it under the term *ozone*, a substance distinguished by its *absence* in the track of plague and pestilence, and by its *presence* in the regions where human health and physical gladness and buoyancy are . . . most indigenous."[17]

The Manchester physician Arthur Ransome (1834–1922) articulated a similar theory about the purifying powers of nature. In an 1876 address, he emphasized the importance of complete oxidation: "We know but little of 'the balancings of the clouds,' but we may be sure that within their recesses all organic matter and impurity is burnt up as certainly as if it had passed through a furnace; the oxygen of the air, intensified in its power by electricity, fastens upon it and changes it into the ultimate products of consumption—carbonic acid, aqueous vapour, ammonia, or nitric acid."[18]

The connection between atmospheric conditions and health fascinated many Victorian researchers. In 1860 the Manchester and Salford Sanitary Association began publishing a weekly report, *Health and Meteorology in Manchester*. In addition to providing such data as temperature, humidity, barometric pressure, wind, precipitation, and cloud cover, the report included measurements taken with an "ozonometer."[19] In an 1866 article in the *Popular Science Review,* the physician and prolific public health expert Benjamin Ward Richardson (1828–96) observed that although ozone had attracted much attention, scientists still knew little about it. When he exposed an eight-year-old flask of rotting ox blood to ozone, he found that its nauseating odor went away. This discovery convinced him that ozone was a powerful purifier that could be used to remove "the putrid emanations of decomposing animal substances." He suggested that this property had commercial possibilities and might someday allow butchers to restore freshness to meat that had spoiled. Although he noted that the precise mechanism for its creation remained a mystery, he declared that ozone was "produced in the atmospheric sea which surrounds our planet." Responding to the widespread view that cities contained too little ozone, Richardson suggested that it might be supplied artificially. Consciously modeling his idea on existing technological systems, he suggested that "an 'ozone company' might prove itself not merely useful but, as a consequence, a paying concern. Such a company could bleach, deodorize, disinfect, preserve meat and vegetables, and give sea air to every household that required it; its 'supply' could be as manageable as gas and as cheap as water."[20]

Discussions of ozone reinforced beliefs about the proper ends of natural processes such as decay and combustion. Both smoke and putrefying matter were understood to consist largely of carbon, and both were believed harmful because they existed in a partially oxidized state. Coal smoke was seen as the result of incomplete combustion, while miasma was seen as the product of incomplete decay. Promoters of ozone argued that it was the perfect antidote to both smoke and miasma, for it—like fire—was thought to destroy impurities by fully oxidizing them.[21] In Ransome's view, the "putrid fermentation" of biological substances released poisonous organic matter into the air. When "well mixed up with it, the air gradually destroys it, burns it up, and turns it into inorganic gases, and it does this with especial rapidity when the air contains that intensely active form of oxygen known by the name of ozone."[22]

Many believed that plants and animals depended on each other not only for oxygen or carbon dioxide, but also for the nutrients that they needed to survive.[23] Human beings and many other creatures ate plants, while plants derived nutrients from manure and decomposing animals.[24] Through both processes, complex animal and vegetable matter could be transformed into simpler substances, eventually culminating in "clean"

Figure 3.1. "A Vision of Utopia" (detail). In this illustration by Sir John Tenniel, science uses ozone to cleanse the air of smoke, fog, and sulphur. From *Punch's Almanack for 1881,* 13 Dec. 1880, n.p.

dirt.[25] This transition was considered essential to life, but it was also thought to be fraught with danger because it produced miasma.[26] Fortunately, nature—given sufficient time and the right conditions—could be counted on to render harmless the sources of miasma. Benjamin Ward Richardson, for example, warned that the improper disposal of dead bodies was dangerous to the community, but noted that if corpses were allowed to decompose in "rapidly-resolving soil . . . the natural cycle of transmutation [would be] harmlessly completed, and the economy of nature conserved."[27] The nature of the risk in both cases was thought to derive from the incomplete oxidation of matter: partial decay in the case of miasma and inefficient combustion in the case of smoke.

Sunlight

During much of the nineteenth century, smoke and sunlight possessed rival claims as disinfectants. As long as disagreement existed, it was difficult to argue that smoke prevention would improve health. This standoff came to an end in the 1870s as a result of bacteriological research. Arthur Downes and Thomas Blunt, experimenting with vials of bacterial solutions, found that sunlight retarded and could even halt the growth of bacteria. Addressing the Royal Society in 1877, they claimed that direct sunlight could completely sterilize solutions containing bacteria, but that smoke-filled or cloudy skies inhibited this effect.[28] According to a contemporary, 75 percent of the light that should have been falling on London was being blocked by smoke.[29]

These new ideas about the health benefits of sunlight and the dangers of obscuring it spread rapidly. As early as 1880 the *Lancet* asserted that fresh air and sunlight were essential for people recovering from respiratory diseases such as tuberculosis, bronchitis, and pneumonia, which together caused over a quarter of all deaths in late nineteenth-century Britain. Because London and other smoky cities were particularly deficient in fresh air and sunlight during cold weather, when people used large quantities of coal for heating, "physicians are obliged to forbid patients to come to town in the winter if there is the slightest suspicion of lung disease." The *Lancet* further asserted that sunlight was valuable in preventing "anemia, rickets, and . . . general feebleness of the children of the London poor."[30] Medical researchers in the early twentieth century reinforced ideas about the health benefits of sunlight with the discovery that the body uses the sun's ultraviolet rays to produce vitamin D. People who never go outdoors or who live in extremely smoky cities may develop rickets as a result of inadequate exposure to sunlight.[31]

In contrast to earlier ideas that traced polluted air to miasma and cautioned those with respiratory ailments to avoid rural areas, keep a safe distance from houseplants, and even inhale coal smoke, during the late nineteenth century many came to see smoke as pollution and to believe that nature was essential to health. As one medical text explained, sanatoria for the treatment of tuberculosis must "have a pure air, free from dust and smoke and the impurities which are inseparable from a dense population." Its author also noted that patients ought to spend as much time as possible out of doors, "since even the best-ventilated room is inferior to the atmosphere of a garden." Pine forests were thought to provide the best surroundings because they supposedly produced large quantities of ozone.[32] Contrary to earlier injunctions to avoid cold and rain, physicians now encouraged their patients to spend their days outdoors and to sleep with their windows open, even in the middle of winter. One tuberculosis patient recalled that when his doctor observed snow blowing through an open window, he smiled approvingly and said, "Wouldn't that be considered suicidal madness under the old *régime?*" By the end of March the patient boasted that he had become as sunburned as he usually was by September.[33] Expressing similar ideas, the president of the Society of Medical Officers of Health asserted that nothing, including drug therapy, was "equal in the treatment of tubercle to bathing the lungs and skin in fresh air." If the air of cities could be freed of smoke and other impurities, he argued, "the death-roll from phthisis [pulmonary tuberculosis] will be so insignificant as to render special sanatoria unnecessary."[34] Two decades later, the antismoke activist Percy Alden declared that the power of smoke to block the health-giving ultraviolet rays of the sun was potentially more harmful than its direct effects on the respiratory system. As he put it in a popular periodical, "The most important diseases affected by polluted air, namely rickets and tuberculosis, are both dependent on its power of reducing the energy of sunlight rather than on its chemical composition." He went on to claim that although the tuberculosis bacillus would remain virulent for six months if kept in the dark, it could be rendered harmless "by a few minutes' exposure to direct sunlight."[35]

One of the early converts to the notion that smoke was unhealthy in large part because it blocked the cleansing rays of the sun was Harvey Littlejohn, the medical officer of health for Sheffield. In an 1897 report on smoke, he noted that

> [d]uring the last few years another important action of light [in addition to contributing to "mental and physical energy"] . . . has been demonstrated, viz., the powerful destructive influence sunlight has upon various forms of bacteria and disease germs, and in the case of one form in particular, the tubercle bacillus, which is the cause of a disease of great prevalence and fatality, it is shown that this organism succumbs after a short exposure to light. Organic changes, or oxidations, which are impossible in darkness readily take place in sunlight.[36]

In 1896 the artist William Blake Richmond told a friend that he had been ordered by his doctors to "go towards the sun for healing." They had

promised him that a period of two or three months in Egypt would restore the "fresh vigour and the old energy that for the moment have left
me."[37] This advice to seek more sunlight had beneficial consequences not
only for Richmond, but for millions of others. Convinced that it was not
enough to seek temporary respite from smoke, he became determined to
eliminate smoke from the air. Two years after his trip to Egypt, Richmond
established the Coal Smoke Abatement Society (discussed in chapter 7).
This group, now known as the National Society for Clean Air and Environmental Protection, is the world's oldest continuously operating antipollution group.

Smog

During the final decades of the nineteenth century, London became synonymous with fog. One observer, writing in January 1880, noted that "in
the dense fog of to-night . . . there are parts of the Strand where the form
of the policeman can be dimly traced at the distance of a yard or two . . . ,
[and] lighted carriages follow torch-bearers slowly across the impenetrable
darkness of Ludgate-circus."[38] In addition to causing inconvenience and
heightening the risk of accidents and crime, these fogs were blamed for
killing thousands of Londoners. Even a day or two of thick, smoky fog was
enough to increase the weekly death rate; episodes of severe fog, such as
those that blanketed London every few years, were blamed for sending hundreds and sometimes even thousands of people to an early grave. After a
dense fog paralyzed London in 1873 and the death rate increased sharply,
the *Lancet* declared that the fatalities were caused by chemicals contained
in the fog, not by the cold temperatures that accompanied it. The journal
pointed out that although the death rate in London rose by 41 percent, the
corresponding increase in England's other large towns was only 8 percent,
despite their equally chilly weather. Declaring this fog to be the most persistent in living memory, it reported that London's air had been nothing
short of "abominable, so laden was it with smoke and dirt."[39] In late January and early February 1880 the mortality rate more than doubled, leading to approximately three thousand "excess deaths" compared to the
same period of the previous year. Many of these deaths were linked to
breathing difficulties; during the week of heaviest fog the number of fatalities attributed to bronchitis was more than four times the ten-year average.[40] Reporting on these deaths, the head of the government statistics office
likened the fog to a disease epidemic. To reinforce this message he observed

Figure 3.2. "'Old King Coal' and the Fog Demon," by Sir John Tenniel. With a smokestack for a crown, a stove for a throne, and an urn spewing carbonic acid (carbon dioxide), the jolly figure of Old King Coal is portrayed as an accomplice to the Angel of Death. Embedded in the dark fog that permeates London are a range of respiratory ailments, including catarrh (inflammation of the nose and throat), bronchitis, pneumonia, and pleurisy. From *Punch,* 13 Nov. 1880, 222–23.

that during the foggy weather the weekly death rate had soared to heights not seen since the worst cholera outbreak of the century.[41]

In *Our Mutual Friend* (1864), Charles Dickens asserted that town fog was of a different character from country fog. In rural areas "fog was grey, whereas in London it was, at about the boundary line, dark yellow, and a little within it brown, and then browner, and then browner, until at the heart of the City . . . it was rusty-black."[42] London's fog was no longer understood as natural, but was instead viewed as an amalgam of nature and culture. Many scientists shared Dickens's view that urban fogs differed markedly from those in rural areas. Louis Parkes, Chelsea's medical officer of health, declared that it was a mistake to assume that London's fog was the result of the geography and meteorology of the metropolis. "Yellow fogs," he told members of the Sanitary Institute in 1892, "are the product of coal combustion mixed up with nature's white mists, the latter being of a comparatively harmless kind."[43] Emphasizing that the fog problem was almost entirely the result of bituminous coal, the meteorologist Francis Albert Rollo Russell (1849–1914) asserted that "all large towns which burn smoky coal have an excess of fogs and darkness."[44] In contrast to the earlier concern that rural fogs were a source of disease, by the late nineteenth century they came to be seen as harmless compared to those in large cities. As Robert H. Scott put it in 1893, "innocent country fogs" of previous years were giving way to those of an "urban character."[45]

Although many people believed that fogs were becoming more frequent and intense in the late nineteenth century, others were skeptical. In a discussion that followed a lecture on fogs at the Society of Arts in 1880, one member of the audience asserted that although "it was assumed that fogs were increasing," this view "was only another illustration of what was often found, that present evils always seemed bigger than past ones."[46] Historical weather records failed to resolve the question, for they often contained subjective distinctions between fog, mist, and haze. In addition, most of these records provided no information about the duration of a particular fog. Striking evidence of the division of contemporary scientific opinion can be found in two papers presented in the early 1890s before the Royal Meteorological Society. Although both Frederick Brodie and Robert Scott relied on the *Daily Weather Report* for their data, they reached opposite conclusions. Whereas Brodie found "a very decided rise" in the prevalence of fog between 1871 and 1890,[47] Scott found "no trace of a regular increase." The president of the society closed the meeting by asserting that even if it could not be shown conclusively that London fogs were greater in number than in the past, it was vital to understand that they were "blacker and more harmful" than previously.[48]

By the end of the nineteenth century, most of Britain's urban inhabitants came to think of fog not simply as aerosolized water vapor, but also as the coal smoke and acidic gases that frequently accompanied it. Indeed, so close was the association between fog and smoke that urban residents often referred to their combination simply as fog. Paradoxically, this practice had the effect of naturalizing smoke, for it implied that it, like the weather, was beyond human control.[49] On Christmas Day 1904, in an attempt to make people recognize that smoke-filled fogs were neither natural nor inevitable, the physician (and treasurer of the Coal Smoke Abatement Society) H. A. Des Voeux (1861–1942) wrote a letter to the editor of the London *Times*. In it, he argued that there was a vast difference between "true fog, or condensation of moisture in the atmosphere" and the smoke-filled fog often found in London. To avoid confusion, Des Voeux suggested the use of a new term, "smog."[50]

CHAPTER 4

The Balance of Nature

The cycle of natural actions and their sequence is regular and
perfect. Interference is dangerous. Our methods of using
carbonaceous fuel are particularly so.
—*B. H. Thwaite, 1892*[1]

WRITING IN 1772, the Unitarian minister Richard Price asserted that numerous processes, most of them biological, depleted the air of its healthy components, but that other biological processes undid this damage. As he put it, "There must be causes in nature, continually operating, which restore the air" when it becomes spoiled.[2] Price's understanding of the workings of the natural world was part of an intellectual framework that the environmental historian Donald Worster has called "the economy of nature." During the eighteenth century, argues Worster, many in both Britain and America began to consider "all of the living organisms of the earth as an interacting whole."[3] Ideas about the interdependence of nature grew stronger in the nineteenth century, thanks in part to the German chemist Justus von Liebig (1803–73), who suggested that human beings and other animals inhaled oxygen and exhaled carbon dioxide, while vegetation took in carbon dioxide and released oxygen.[4] Viewing this symbiotic relationship through the lens of natural theology, the physician S. Scott Alison asserted in 1839 that "a wise and good Creator . . . has so ordered it that one department of nature shall correct the bad tendencies of the other." God, he explained, "has placed a weight at the opposite end of the balance to counterpoise and balance the glorious work of his hand. Animal life is met by vegetable life: their results are made to neutralize those of each other."[5]

Despite clear evidence that human beings could alter natural conditions on a local scale by damming rivers, digging canals, and turning fields into cities, most believed that such actions could have no effect on the larger environment. Articulating this view in 1866, the eminent physician Benjamin Ward Richardson asserted that "whatever man may do, he cannot disturb the motion of the earth, nor change its balance a single grain. Nature, Almighty conservator! from age to age ever keeps up a continuous supply for demand."[6] Reinforcing Richardson's position, the Freiberg chemistry professor Clemens Winkler (1838–1904), who later discovered the element germanium, argued in 1877 that "man's hand is too weak to interfere noticeably with the imposing mechanism of the cosmic gear. We work on a small scale and too slowly to disturb the equilibrium of the proportions [of oxygen and carbon dioxide] ruling on earth."[7]

As people gained increasing power over the environment, however, some began to express concerns about the consequences. One of the first to do so was the scholar and diplomat George Perkins Marsh (1801–82). In *Man and Nature; or, Physical Geography as Modified by Human Action* (1864), Marsh challenged the assumption that the environment was so resilient that human beings could not measurably alter it. During the last quarter of the nineteenth century, the atmosphere became an important focus of debates about humanity's effect on the natural environment. In 1895 C. M. Aikman directly contradicted Winkler's earlier calculations and declared that it was likely that "the amount of carbonic acid gas in the atmosphere is slowly increasing."[8] One year later, unleashing a debate over climate change that continues to this day, the Swedish chemist Svante Arrhenius (1859–1927) suggested that such an increase could lead to global warming.[9]

Respiration and Combustion

Long before people began to worry about excessive levels of carbon dioxide on a global scale, they expressed concerns about its consequences within confined areas. During the eighteenth century, many people attributed the high prevalence of disease aboard ships and in jails not only to miasma, but also to insufficient quantities of "respirable air" (eventually identified as oxygen).[10] Such beliefs were reinforced during the nineteenth century when health experts such as Thomas Southwood Smith declared that similar conditions existed within crowded apartments, theaters, and lecture halls.[11]

During the nineteenth century, some began to worry that entire cities were becoming dangerously deficient in fresh air. These concerns grew as Britain became increasingly urban. The 1851 census revealed that more people in England and Wales lived in towns and cities than in rural areas. (Seventy years would pass before the United States reached this level of urbanization.) In 1879 the medical professor F. S. B. François de Chaumont pointed out that London's four million human inhabitants, its army of horses, and its countless fires transformed vast quantities of oxygen into carbon dioxide.[12] Without a constant influx of rural air, the inhabitants of cities would rapidly suffer the same fate as mice trapped with burning candles in sealed jars. According to some, this was already starting to occur. In 1894 Eric Stuart Bruce, a fellow of the Royal Meteorological Society, stated that carbon dioxide acted as "a virulent poison when breathed into the system," and declared that it was the most hazardous substance found in urban air. In his view, the languor and headaches associated with periods of dense fog indicated that its sufferers were being "partially asphyxiated" by carbon dioxide.[13]

To prevent this from happening, a wide spectrum of public health experts, social reformers, and conservationists sought to increase the number of oxygen-producing plants and trees in Britain's largest cities.[14] In 1877 the public health expert Alfred Carpenter urged city dwellers to "restore the proper equilibrium in the constituents of the air; and to get rid of the unnatural elements and bring back those which are missing."[15] Leaders of the Commons Preservation Society (established in 1865) declared that their efforts were designed as much to improve public health as to save aesthetically pleasing places.[16] Parks within London were becoming increasingly necessary, many argued, because urban growth was pushing the countryside ever more distant and lessening the likelihood that breezes would bring in unpolluted air.[17] The issue of urban green space even occupied an important role in the politics of the London County Council during the 1890s.[18] As part of its 1896 campaign efforts, the London Liberal and Reform Union issued a brochure entitled *London's Lungs,* which argued that the public acquisition of parks was well worth the modest increase in property taxes needed to pay for these "additional breathing spaces."[19]

Concerns about the effects of respiration on oxygen levels, while alarming to some, were dwarfed by anxieties about combustion. In the 1890s Londoners consumed about 12 million tons of coal each year. According to a tract published by the Society for Promoting Christian Knowledge in 1895, burning just 25 pounds of coal produced as much carbon dioxide in two hours as the respiration of 1,000 people.[20] Based on this estimate, the carbon dioxide produced by London's coal combustion was equivalent to the exhalation of an additional 100 million human beings. Of course, coal use varied enormously between different seasons and times of day. On a cold winter's morning, the amount of carbon dioxide coming from London's fireplaces and stoves may have exceeded that from the lungs of every person then alive in the entire world.[21]

Coal fires produced more than just carbon dioxide, however. In contrast to respiration, which was advantageous to vegetation, coal combustion appeared to violate "the economy of nature" by interfering with its ability to restore the air to health. As one writer put it in 1892, if coal were burned perfectly, "the carbon dioxide produced would be available for absorption by vegetables; but our methods are so imperfect that, instead of producing carbon dioxide, we send into the pure air volumes of sooty particles which are poisonous to animals and to vegetables alike."[22] Incomplete combustion, instead of producing essential carbon dioxide, yielded smoke and soot that blocked the pores of leaves and interfered with their ability to absorb carbon dioxide and produce the oxygen that people needed to breathe. Foreshadowing environmentalists a century later, the Glasgow medical officer of health James Burn Russell (1837–1904) warned in 1876 that humanity was dangerously upsetting "the cycle of interchange and co-operation of the animal and vegetable kingdoms" on which life depended.[23] If the ability of plants to create oxygen were further damaged, warned the Lancashire coal mine owner Herbert Fletcher in 1888, the level of carbon dioxide would rise to unhealthy levels.[24]

During the closing decades of the nineteenth century people across the political spectrum asserted that efforts to increase the amount of "nature" in urban areas were being hampered by the smoky air that enveloped and damaged the trees and plants they were trying to protect. Parks, instead of being places of beauty, were being "rendered hideous by the blackness of everything within them—trees stunted, dying—flowers struggling to bloom, and sometimes their species barely recognizable."[25] Given the presumed interdependence of plants and animals, devastation to one was seen as a serious threat to the other. For some, the desire to make cities greener led directly to efforts to reduce smoke. Others suggested a less ambitious solution: If the air of London made it impossible to grow healthy lime trees and English yews, they should not be planted there. More hardy was the tulip tree, which was recommended as an "excellent subject for town planting." No matter how "scorched, blackened, and encrusted with soot as the foliage may appear at the end of the summer," noted the plant expert A. D. Webster, "the following spring it again puts forth a garb of the freshest and healthiest greenery."[26] The London plane, even more resistant to soot and smoke because it periodically shed its bark, was also widely recommended.[27]

Circulation

Dismissing efforts to increase the production of oxygen in cities, some maintained that respiration alone consumed far more oxygen than could ever be provided by vegetation within cities. One prominent medical expert claimed in 1893 that a single human being depended on the oxygen produced by one acre of vegetation. No city of any size, he pointed out, could survive without importing oxygen from a vast hinterland.[28] The engineer and ventilation expert Douglas Galton (1822–99)—a first cousin of the eugenicist Francis Galton—embraced this view. He believed that air, like water, would become hazardous if stagnant. Without a constant influx of oxygen-rich air from the country, urban dwellers would quickly become poisoned.[29] To prevent this from occurring many called for wider streets, which would promote not only the circulation of traffic but also the movement of air. Inspired by the wide and straight boulevards that Baron von Haussmann had recently carved through Paris, Galton claimed that "if we could cut London in two from north to south by a broad open space, as it is cut in two from east to west by its river, it would do something to break up the cloud of smoke."[30]

In contrast to those who were confident that the natural circulation of the air was impervious to human intervention, others worried that it might be interrupted with deadly effect. Such fears surfaced in both popular and scientific writings. In an 1884 address to the Sanitary Institute, John Collins claimed that the dangers of impure water were minor compared to those of impure air, because "we do not drink air as we drink water—intermittently—and to a comparatively small extent. We imbibe it equally during our sleeping as in our wakeful hours, and the *smallest* interference with the regularity of this supply is fatal."[31] Eight years later, in 1892, a short work of fiction appeared in the *Idler* magazine. Entitled "The Doom of London," the story described a catastrophic fog that suffocated all but a handful of the city's inhabitants.[32] In the wake of this story's publication, several suggested that its premise was plausible. According to an 1893 article in the *Gentleman's Magazine,* if a smoke-filled fog persisted long enough, "the whole population might be poisoned by the carbonic acid [carbon dioxide] with which the air would in that space of time be saturated."[33]

Dismissing natural forces as inadequate to the task of ventilating a great city, some argued for a mechanical solution. Following the same approach that he had championed in supplying urban residents with pure water,[34] Edwin Chadwick, along with his friend, Dr. Neil Arnott, discussed establishing "a pure air company." Prefiguring oxygen bars by more than a century, the business that they contemplated in the 1880s was to "draw the air from a suitable height above the common layers, and distribute it into houses by engine power, or, as gas is distributed, and . . . give even a better air than people generally obtain in suburban residences."[35]

The Wider Environment

Even before the massive engineering effort associated with the building of railroads across Britain, it was clear that human activities could radically reshape the natural world. One of the most conspicuous examples of this phenomenon was urban growth, memorably depicted in "London Going out of Town; or, The March of Bricks and Mortar" (1829). In this drawing, George Cruikshank suggested that the expansion of London into the surrounding countryside was an invasion that uprooted trees, tore up meadows—and filled the air with dense clouds of coal smoke.[36] Despite those such as Cruikshank and Wordsworth who criticized humanity's impact on the natural world in the early nineteenth century, most of their contemporaries equated it with progress, both economic and moral.

During the last quarter of the nineteenth century, however, coal smoke became increasingly associated with violations of the natural order. One who held this view was the four-time Liberal prime minister William Ewart Gladstone (1809–98). In 1877, during a period when the Conservatives held power, Gladstone decried the pollution that filled Britain's cities. In contrast to those who justified environmental degradation by arguing that they were simply following God's injunction to exercise dominion over nature (and who maintained that the natural environment would in any case quickly rebound if harmed), he used equally religious language to argue for what has since come to be called the stewardship interpretation.[37] "God," declared Gladstone, "made this world to be pleasant to dwell in. I don't mean to say He meant it to be without toil and affliction, but He made our natural and physical condition to be pleasant. The air, and the sun, and the skies, and the trees, and the grass, and the rivers, they are pleasant things, and we go about spoiling and defacing and deforming them."[38]

Gladstone's critique of air pollution resonated with many of his contemporaries. Speaking to an antipollution gathering in Manchester in the 1880s, the chemist J. Carter Bell compared the blind pursuit of technological progress with the self-destructive quest for forbidden knowledge that had led to Adam and Eve's exile from the Garden of Eden. "Science has made great strides," he noted, but "the result has been in many instances to convert the garden into the wilderness." Continuing in this biblical vein, he described industrial effluents as "dews of hell" that stripped trees of their leaves, converted watercourses into "Stygian streams" of acid, and produced a landscape that was the "very picture of desolation."[39]

Cities were not inherently unhealthy, argued many reformers, but they became that way through human actions, particularly coal smoke. As the physician Alfred Maddock put it in 1860, "The natural salubrity of London is self-evident. The gently sloping position, its gravely beds, its copious springs, its squares, parks, and open places, its tidal river, the breezy hills by which it is encircled on nearly every side, all prominently proclaim that it ought to be, *par excellence,* the chosen abode of health."[40] In his opinion human choices had befouled the air of London in violation of divine will. Over two decades later, the medical officer of health John Leigh made the same point in reference to Manchester. He argued that its pure water, good diet, high wages, and excellent sewerage and drainage ought to have made the city "a model of urban health." Instead, the large quantity of coal smoke in the air obviated all these beneficial qualities and made the Manchester worker "one of the most unhealthy men in the kingdom."[41]

In contrast to those who assumed that cities were inevitably unhealthy places, others argued that they need not be this way. For although human actions had despoiled cities, so too could human actions restore them to purity. In *The Earthly Paradise* (1868–70), the socialist artist and writer William Morris (1834–96) called on his readers to

> Forget six counties overhung with smoke,
> Forget the snorting steam and piston stroke,
> Forget the spreading of the hideous town;

> Think rather of the pack-horse on the down,
> And dream of London, small, and white, and clean,
> The clear Thames bordered by its gardens green.[42]

The problem of pollution-damaged plants was not limited to cities. According to one commentator, much of northern England lay beneath a thick cloud of smoke that was harmful to vegetation.[43] Foreshadowing later concerns about human degeneration, one Lancashire landlord declared in 1881 that as a result of smoke, "Trees which I have planted more than 20 years ago, after struggling through a sickly youth, begin to decay before they have reached anything like maturity."[44]

One of the most surprising effects of smoke on the natural environment was that it was causing a decline in certain varieties of moths and an increase in others. This development intrigued many late nineteenth-century scientists and helped to awaken in them an awareness of the ecological impact of air pollution. Emphasizing the significance of this phenomenon, the biologist Nicholas Cooke suggested in 1877 that human action was transforming nature to such an extent that it was shaping the course of biological evolution and "carrying into effect the laws of creation before our eyes."[45] Contemporary entomologists disagreed as to why black moths were becoming more common, but many believed that "progressive melanism" was related in some way to the growth of industry and cities. The first black-colored example of the peppered moth was captured in Manchester in 1848; others were soon found throughout Lancashire, Yorkshire, the north Midlands, and in London. By the 1870s light-colored varieties were virtually nonexistent in many smoky areas.[46] Although it was widely known that cities, particularly industrial ones, were centers of moth melanism, contemporaries disagreed about why it was occurring.[47] Some attributed it to the soot that coated the plants which these moths ate, while others suggested (correctly, as it turned out) that dark moths were better able to conceal themselves from birds against smoke-darkened lichen than their lighter cousins and were thus more likely to survive and reproduce.[48] As late as 1945 hardly any light-colored examples of the peppered moth existed in smoky cities like London.[49] By the 1980s, however, light-colored strains had reappeared in many parts of Britain.[50]

Ubiquitous air pollution also threatened Britain's ability to feed itself, warned the German-born chemist Augustus Voelcker (1822–84). In addition to being affected by natural influences such as rainfall, wind, and temperature, crops were being damaged by acids that mimicked the effects

Figure 4.1. "Our Smoky River!" This cartoon by Linley Sambourne contrasts the allegedly pure air and water of Pepys's London with the murky conditions found in Mr. Punch's London three centuries later. Ironically, the year depicted on the left—1661—was the same year in which John Evelyn published his diatribe against smoke in London. From *Punch,* 24 Sept. 1898, 134.

of drought or frost.[51] The combustion of coal filled the air with massive quantities of sulphur dioxide—75 million pounds a year in London alone, according to an estimate in 1880. When this compound mixed with moisture in the air, the result was highly corrosive sulphuric acid, a major source of acid rain.[52]

Damage to crops appeared all the more serious because Britain did not produce enough food to feed its people. An 1877 article in the *Times* had warned that dependence on foreign grain made Britain vulnerable to food shortages in wartime—a prediction that later became a reality during both world wars. Britain's dependence on imported grain quadrupled between 1850 and 1880 as domestic wheat production declined and the country's population grew. Wheat production fell even further in subsequent years, supplanted by low-cost imported grain from the Americas. As the century ended, roughly three-fourths of the wheat consumed in

Britain came from abroad.[53] London, which produced no food save a tiny quantity of garden vegetables, was especially vulnerable. If a dense fog lasted for a week without interruption, warned some, food might fail to reach the metropolis.[54]

By the late nineteenth century the belief that illness resulted when people lost touch with nature was extremely widespread. "Steam and free trade," argued the medical professor George Vivian Poore (1843–1904), had led people to forget that their existence depended on natural processes. "We have neglected the earth unit absolutely, and we are encountering serious difficulties in consequence, and one of these is the thick impure atmosphere of cities, and the dense black fogs when the climatic conditions are favourable for their formation."[55] Rejecting the assumption that the Earth had an infinite capacity to absorb and purify the by-products of industrial society, a growing number of people began to agree with the Manchester physician Arthur Ransome, who warned that nature was vulnerable, and that human interference had outpaced the ability of the environment to compensate for the degradation that people caused. Although maintaining that winds, storms, rain, diffusion, and vegetation did much to purify foul air, he asserted that the cleansing powers of nature were "not equal to any burden that may be laid upon them. The destruction of plant-life, and the injury to human beings that occur where the air is loaded with these substances, show that something more is needed than Nature unaided is able to perform." In his view, sinful men had corrupted nature, a defenseless female who depended on virtuous men for her rescue: "She cannot stifle the stokers of furnace fires when they carelessly allow volumes of black smoke to escape. . . . She cannot turn the stream of evil vapours and gases back into the works from which they pour. Legislation and vigilant inspection can alone interfere with this."[56]

CHAPTER 5

Pollution and Civilization

To pile up 100,000 factory chimneys, vomiting soot, to fill the air with poisonous vapours till every leaf within ten miles is withered, to choke up rivers with putrid refuse, to turn tracts as big and once as lovely as the New Forest into arid, noisome wastes; . . . and overhead by day and by night a murky pall of smoke—all this is not an heroic achievement.
—*Frederic Harrison, 1882*[1]

Air in such a country as this with no part of it sixty miles from the ocean would always be fresh and clean were it not for the uncivilized people of high cultivation and education who inhabit these islands.
—*H. A. Des Voeux, 1936*[2]

IF MOST OF THE Victorian age was characterized by an optimism about technology and a faith in progress, the *fin de siècle* saw the emergence of widespread anxiety about the future on many levels: economic, political, military, social, and even biological. This anxiety was expressed in many ways, among which attitudes toward urban and industrial society assumed an important place. The great paradox of modern life, in the eyes of many pessimists, was that "progress" seemed to be undermining civilization. As the rate of change increased, society appeared to become less stable socially, culturally, and environmentally. Coal smoke was central to such concerns, and it became a powerful symbol for them.

Coal use on the scale prevailing in nineteenth-century Britain depended on an interlocking network of workers, technology, and economics that carried coal from deep underground to the millions of fireplaces, stoves, and furnaces where it was consumed. Many observers feared, however, that the smoke that resulted from this reliance on coal undermined the complex civilization that made its consumption possible. Herbert

Philips (1834–1905), a prominent reformer in Manchester, declared in 1876 that if nothing were done to reduce the pollution of its air, by the twentieth century everyone who could manage to leave the city would flee, a situation that "no one could consider . . . a triumphant result of what we called modern progress and Civilisation."[3]

Nineteenth-century commentators frequently viewed damage to the environment as an indictment of modern life. As urbanization increased and society experienced rapid change, many people looked to the countryside as a source of stability. Yet two things complicated the process. First, it became increasingly difficult to reconcile Britain's predominantly urban society with a national identity rooted in the countryside. Second, the distinction between urban and rural became blurred as a result of developments in transportation and communication, and by discoveries that no part of Britain was untouched by the smoke of its large cities. Reformers argued that a society that poisoned trees and shrubs, blocked the sun's rays, made agriculture unprofitable, and disfigured both built and natural landscapes, was in trouble.

Faith in progress, whether economic, technological, or moral, coexisted in the late nineteenth century with a growing ambivalence about the long-term sustainability of modern society. One manifestation of this anxiety was the theme of urban apocalypse that occupied a central place in late nineteenth-century fiction. In 1886 the nature writer Richard Jefferies published *After London, or Wild England,* a portrayal of the great metropolis "deserted and utterly extinct." Jefferies' depiction of a journey into a future London resembles other contemporary narratives that attracted middle-class readers with glimpses of sites of danger and fascination, such as the African interior and the slums of London. Although the nightmarish descriptions found in the book were ostensibly fictional, they may be read as a parable for what had already begun. Prefiguring Rachel Carson's *Silent Spring* by three-quarters of a century, the book portrays a world in which

> the earth was poison, the water poison, the air poison, the very light of heaven, falling through such an atmosphere, poison. There were said to be places where the earth was on fire and belched forth poisonous fumes, supposed to be from the combustion of the enormous stores of strange and unknown chemicals collected by the wonderful people of those times.[4]

Many people believed that the environmental character of a place—which they believed could be discerned readily by its smell and appearance—exerted a powerful effect not only on health but also on morality and behavior. The countryside was often portrayed as a haven of order, tradition, and stable class relations, while cities were depicted as chaotic, rootless, and anarchic. Speaking in Manchester in the mid-1880s, the Lancashire botanist Robert Holland asserted that nature helped to maintain social control. Interspersed among his claims about the ennobling effect of nature were more pragmatic assertions that periodic excursions to the countryside made workers "better in every way—better in health, elevated in mind, more helpful, and better fitted to go back to their work." If it were not possible for workers to escape into the country, argued Holland, parks should be created in the middle of cities to bring "the country . . . to the very doors of our working classes."[5] At the end of the nineteenth century the artist and antismoke leader William Blake Richmond echoed the notion that environmental changes could mitigate class conflict. Once the air was free from smoke, he asserted, poor people would "lose their sullen looks and become more bright and cheerful."[6] William Booth, the founder of the Salvation Army, shared this perspective and blamed the "foul and poisoned air" of large cities for contributing to drunkenness. Men drank, he argued, to compensate for the inadequate supply of vitalizing oxygen and ozone in the air of cities.[7]

Victorians had mixed feelings about the increasing separation of the classes. On the one hand, many members of the middle classes wanted to live as far as possible from poor and working-class neighborhoods. On the other hand, they feared that the growing spatial gap between rich and poor was dangerous. In 1855 *Chambers's Journal* asserted that "the smoke-charged atmosphere" of cities "leads to moral results of a most unfortunate kind, in as far as it sends away the rich to dwell apart from the poor, who are thus deprived of the neighbourly sympathy . . . and edification which they might otherwise obtain from their more fortunate and better-educated brethren."[8] Commenting on the same phenomenon a generation later, another observer explained that "the wealthiest people are always in the advance," moving ever upwind from the smells and smoke of cities.[9] As a result of prevailing winds, places in east and southeast London, such as Poplar, Islington, Southwark, and Woolwich, contained noticeably more smoke than did the West End.[10] Many reformers, whether involved in the settlement-house movement, housing reform, or efforts to counter the growth of air pollution, believed that social stability depended on the dissemination of middle-class values among the lower orders, and feared that the weakening of such contact might lead to the formation of aggressive class consciousness among the masses.[11]

Others shared this concern. In 1877 a "deputation of gentlemen" presented a memorial to the lord mayor of Manchester that called for him to do more to control smoke. They argued that "the well-to-do inhabitants" were moving out of the city, leading to "apathy on their part towards the condition of the lower classes. Indeed, the depressing effects of our impure air may fairly be considered as a powerful factor in retarding efforts at social reform."[12] Discussing this phenomenon in a lecture delivered in Manchester in the 1880s, Robert Holland noted that "[t]he rich can leave the sordid city and make their homes in the beautiful country far away from their business; the poor cannot do so. They must breathe the stifling, smoky atmosphere from one year's end to another." This resulting class segregation, Holland warned, was leading the rich to become callous toward those less fortunate than themselves, and the poor to respond with "a feeling of envy and enmity."[13] Arthur Ransome, a fellow of the prestigious Royal College of Physicians of London, asserted that smoke discouraged people of different classes from living "side by side as they used to do."[14] When this occurred, poor people and criminals (often conflated in contemporary thought) escaped the watchful eyes and moral influence of their social superiors. Middle-class representations of working-class Londoners, particularly those who were recent immigrants from Ireland or southern or eastern Europe, were strikingly similar to Victorian stereotypes about Africans. Late Victorian writers routinely portrayed both groups as dangerous, disorderly, unhealthy, and exotic. William Booth's use in 1890 of the phrase "darkest England" no doubt struck many of his contemporaries as a provocative and obvious analogy to "darkest Africa."[15]

In a series of articles published in 1883, journalist George Sims decried the environmental conditions found in the slums of Britain's large cities. Sims declared that "in terms of water and air the most degraded savage . . . is a thousand times better off than the London labourer and his family." Pointing to spheres where compulsion had already been accepted, Sims advocated greater state intervention to improve the urban environment: "The law says that no child shall grow up without reading, writing, and arithmetic; but the law does nothing that children may have air and light and shelter." In contrast to those who treated air pollution principally as an aesthetic problem, Sims called on his readers to recognize that it was responsible for "the wholesale stifling and poisoning of the poor which now goes on all over London."[16] The novelist and Fabian H. G. Wells expressed similar concerns a decade later in *The Time Machine* (1895), in which his narrator asks, "Does not an East-end worker live in such artificial conditions as practically to be cut off from the natural surface of the earth?"[17]

Implicit in many discussions of the urban environment was the assumption that individuals of the middle and upper classes were able to appreciate fine surroundings, but that they would be wasted on the working class, which supposedly lacked discernment. Writing in 1828, Daniel Ellis argued that "smoke, smells . . . [and] effluvia" encouraged everyone who could manage to do so to move away. "Those whose business or whose property oblige them to remain are compelled to live on in discomfort and disgust, to see their property daily deteriorating in value—their old neighbours driven away—and a new and inferior class of people settling around and among them."[18] In the 1840s Friedrich Engels similarly observed that the "eastern and north-eastern districts of Manchester are the only ones in which the middle classes have not built any houses for themselves. This is because for ten or eleven months in the year the winds blowing from the west and southwest always carry the smoke from the factories—and there is plenty of it—over this part of the town. The workers alone can breathe this polluted atmosphere."[19] Instead of realizing that economic necessity constrained the choices that many people could make about where to live, nineteenth-century observers frequently implied that poor people lived in ugly and unsanitary environments because they lacked the desire for something better. Thus, conditions that it would be unreasonable to expect a person of "quality" to tolerate could be justified as perfectly acceptable for the lower orders.

Scarcity

Another source of concern sprang from fears that Britain was rapidly exhausting the coal that was fundamental to its industrial and imperial preeminence. In response to Jevons's alarming predictions of imminent crisis, a government commission in the early 1870s looked into the matter and concluded that Britain's coal measures would last for at least 276 years and perhaps for as much as a thousand years more than that.[20] Despite such reassurances, concerns about the finite nature of the coal supply served as a surrogate for fears about the apparent passing of Britain's industrial

Figure 5.1. "The New Zealander" (detail). In this engraving by Gustave Doré, an inhabitant of Britain's colonial empire contemplates the future ruins of imperial London. The image, inspired by the historian Thomas Babington Macaulay, struck a nerve among people who feared (or hoped) that Britain's power would not last. From Gustave Doré and Blanchard Jerrold, *London: A Pilgrimage* (1872), facing 188.

preeminence. In 1885 the chemist Thomas Fletcher suggested that when the English ran out of coal, they would be surpassed by their colonial subjects, who would someday visit the ruins of imperial London and walk among the crumbled stones of Westminster Abbey.[21]

Professor Leonhard Sohncke similarly held that the exhaustion of Britain's coal reserves would bring about the end of her status as the world's leading power. Arguing that virtually "the entire wealth of England" depended upon the nation's "treasures of coal," he warned in 1890 that it was a nonrenewable resource: "Coal represents for us a certain amount of capital that is not to be increased, that bears no interest, and that is by no means inexhaustible." Since the invention of the steam engine Britain had been "wasting in the most irresponsible manner our capital of mineral coal that we cannot replace." Once Britain had depleted its reserves, non-European countries would be able to charge high prices for their coal and force Britain into a state of economic dependency. The result, predicted Sohncke, would be the "displacement of the centre of civilization from Europe to these coal countries." To prevent such a calamity from occurring, Sohncke suggested developing alternative sources of energy, including water and wind power, and the "energy of direct solar radiation." He noted that a parabolic reflector that used concentrated sunlight to boil water had been exhibited at the 1878 Paris Exhibition, and also suggested that sunlight might be converted directly into electricity.[22]

Concerns about the longevity of Britain's coal supplies continued in the early twentieth century. The Quaker writer John W. Graham (1859–1932), who would later chair the Smoke Abatement League of Great Britain, argued in 1907 that Britain's pride in the output of its coal mines was misplaced. Unlike other industries, the coal industry drew from a source that was finite and nonrenewable. Urging his contemporaries to save more coal for the future, he asserted that present profits in the coal industry left Britain "permanently impoverished." As Britain's coal supplies dwindled, her economy would be at the mercy of foreign coal suppliers, and its status as a world power would disappear.

> When our coal has gone, the manufacturing and mercantile part of the greatness of England and all that depends upon it will have gone too. London will live by running hotels in which Americans can spend their holidays, and as a centre of culture and fashion; in Lancashire and Yorkshire sheep will wander over the ruined heaps of former towns; Manchester and Leeds will be visited chiefly for their art galleries and libraries, their impoverished universities and interesting old town halls, doubtless cleaned at last.[23]

Graham's predictions proved surprisingly accurate not only in their forecast of industrial decline but also in regard to the attempts that such industrial cities have made in recent years to promote tourism. Leeds town hall remained stained with a black layer of soot until the early 1970s. As it was being cleaned, many lifelong residents of the city—who had never realized that the building was made of cream-colored stone—thought that it was being painted white.[24]

Decline

If the twentieth century can be called the American Century, the nineteenth was undoubtedly British. Cheap and abundant fossil fuels helped to make both nations into superpowers. Britain became the most powerful economic and political force in the world during the age of coal, and the United States did so during the age of oil. Ever since its oil production peaked a generation ago, the United States has become increasingly dependent on imported oil. Should this supply disappear, some warn that the consequences—both economic and geopolitical—could be catastrophic for the United States. Britain faced similar concerns over a century ago as its coal production declined, at first relative to other countries, and, from 1913, in absolute terms.

Writing in 1879 in the midst of a long-lasting economic recession, the Irish political economist T. E. C. Leslie declared that the United Kingdom's "manufacture and commerce may or may not recover their vigour and supremacy. Our agriculture may or may not be overborne finally by American competition." Leslie described his fears about the future with atmospheric metaphors, referring to a "gloom of uncertainty" and a "sense of being in the dark."[25] Concerns about Britain's economic prospects overlapped with fears of political and social upheaval, and London fog provided an apt symbol for people afraid of what lay ahead. Employing metaphors similar to those used by Leslie, the *Illustrated London News* summed up 1879 as "a year of continuous gloom. Cheerless weather, bad trade, social discomfort, [and] unforeseen political disasters have made up the staple of experience in the United Kingdom, with only here and there a bright interval to relieve it."[26] As foreign competition grew and British manufacturers suffered through a lengthy economic downturn that some referred to as a "great depression," industrialists argued that the smoke they emitted was an indication of jobs and profits. During the late nineteenth century, recalled one antismoke activist, one frequently heard the expression, "There is no brass without muck." According to him, both factory owners and their employees "were pleased to see smoke as a sign of prosperity."[27] Criticizing manufacturers for this attitude, Robert Holland maintained in the 1880s that the true cost of pollution was far greater than people realized. Like environmentalists today who argue that the market fails to account for "externalities" such as pollution, Holland claimed that the damage caused by air pollution from chemical works exceeded the profits of the entire industry.[28] The antismoke campaigner Rollo Russell,

Figure 5.2. "A Second String." In this cartoon by Leonard Raven-Hill, King Coal apologizes to Britannia that he's missed "one or two of our dances." She replies that there is no harm done, for "[i]t's given me the opportunity of making the better acquaintance of Prince Petroleo—very nice and gushing. You mustn't think you'll always be indispensable." From *Punch*, 10 Apr. 1912, 263. Reproduced with permission of Punch Ltd.

son of the Liberal prime minister Lord John Russell, similarly maintained that the costs of repairing the damage that smoke caused to clothing and buildings in London exceeded the amount spent on coal.[29] In 1899 the *Builder*, one of the leading architectural publications of the period, claimed that smoke imposed less obvious costs in addition to those involving increased need for cleaning and lighting. "Damage to gardens and parks, fields and woods," as well as "delay and danger to traffic" could not be measured in "pounds, shillings, and pence," but they were no less real. The major impediment to further progress in reducing smoke, it argued, was the persistent belief among many factory owners that smoke was "a trophy of advancing trade."[30]

In an effort to change the minds of industrialists who complained that a period of low profits and increasing foreign competition was no time to advocate stricter enforcement of antismoke laws, William Bousfield argued in the *Art Journal* in 1882 that "in the war of hostile tariffs which assails our export trade, there is but one way of maintaining our industrial ascendancy, and that is by the excellence and the artistic beauty of our manufactures." Insisting that the design and manufacture of products deteriorated if conducted in an atmosphere of "gloom and ugliness," Bousfield maintained that the smoke that had "arisen in the creation of our trade . . . must be removed if we are to preserve it."[31] Reginald Brabazon, Twelfth Earl of Meath (1841–1929), shared this view. Meath, an Anglo-Irish landowner with a seat in the House of Lords, was an ardent proponent of both imperialism and environmental reform.[32] In 1887 he asserted that "the nation which has the healthiest and sturdiest human material with which to work will produce the best and most saleable manufactures." How could Britain maintain its position in the global economy, he asked, "if the intellect of our designers is weakened by bad health, and the bodies of our artisans and labourers are suffering from lassitude and depression?"[33]

In May 1921 a massive miners' strike disrupted Britain's coal supply to an extent unseen since the English Civil War. While large numbers of people complained of the inconvenience and hardship occasioned by the coal stoppage, many also welcomed its effect on the atmosphere. As Edward Carpenter put it, "The shortage of fuel, the silence of the factories, the emptiness of thousands of domestic grates, are teaching us—even through much suffering—what it means to have a really pure and unpolluted sky."[34] The *Sheffield Daily Telegraph*, which shared much of Carpenter's dislike of smoke, declared that the city "has been a wonderful place to live during these past eight weeks, and if only we could find a way of setting our unemployed to work, we should be willing enough to see the coal stoppage go on for ever."[35] When coal resumed flowing from the mines a month later, the paper expressed the hope that the experience of clean air might inspire greater resolve against smoke: "We have learnt that even in a monstrously-overgrown and hideously-contrived city like Sheffield, life is nevertheless bearable, even enjoyable, when the smoke is absent. If we cannot abolish smoke entirely, we can surely reduce it and so lessen its evil effects."[36] Not everyone agreed with this view of things, however. A letter to the editor argued that the true threat to health came not from industrial smokestacks but from unemployment: "A works producing smoke is a godsend to the town at present, and . . . should be encouraged."[37] The apparent conflict between jobs and the environment is one that reverberates to this day.

Reporting in 1921 on the Whit Monday holiday that had just taken place, the *Times* noted, "Yesterday was sunburn-day in London—the most severe for at least 50 years. For yesterday, thanks to the coal stoppage and the holiday combined, the London atmosphere had become so clear of smoke that the burning rays of the sun—the so-called violet and ultra-violet rays—were no longer filtered out. You could get a good honest sunburn in Bond-street or Piccadilly."[38] Another article in the same issue reported that those who had spent the holiday in the capital had "basked in the brilliant sunshine and marvelled at the clearness of an atmosphere unpolluted by its usual quantum of smoke and grime. The reduction in the coal consumption has had a marked effect in increasing the attractiveness of London's great open spaces."[39] The day also prompted a letter to the editor of the *Times* from a person who tried to convince readers to reject their long-standing acceptance of polluted air as normal and inevitable, and to open their senses to the natural beauty that had been temporally brought to light: "Probably for the first time since coal was generally used, we are enjoying, owing to the coal shortage, a pure atmosphere. The whole aspect of London is changed. Vistas hitherto undreamt of reveal themselves in every direction—a walk in Kensington or St. John's Wood is like a walk in the country. Fresh, clean foliage refreshes the eye, and the scent of lilac and may fills the air. It is depressing to think that from sheer apathy, and because it is nobody's business, we shall go back to the old conditions directly coal is again available."[40] This prediction proved all too accurate. As soon as the strike ended, the air quickly returned to its "normal" smoke-filled state. Even if most people regretted the resumption of smoke, few believed that the government ought to impose limits on the household use of coal. As the *Times* editorialized in the summer of 1921, "We are still far from the point at which compulsion is possible. Public opinion has to be educated."[41]

Less than a decade after the 1921 coal stoppage, economic and social upheaval again resulted in a sharp reduction in coal smoke. Instead of being caused by striking miners, the disappearance of smoke in the early 1930s was due to the paralysis of global capitalism known as the Great Depression. Although the skies of many cities became noticeably cleaner, few workers were inclined to recognize this as a change for the better. In the run-up to the October 1931 general election, the Conservative Party issued a poster that was designed to appeal to voters' concern over lost jobs. Emblazoned in large script across its top were the words, "Smokeless Chimneys and—ANXIOUS MOTHERS!" The background contained the silhouettes of factories idled by the Depression, their smokestacks noticeably free of any smoke. In the foreground stood a distressed woman to whom two young and desperate-looking children were clinging. Between this threesome and the distant factories was a cluster of unemployed working men, one of them presumably the husband and father of the three central figures, unable to provide for his family. The artist heightened the message of male inadequacy by making the unemployed workers appear small and anonymous. In contrast to the mother and her children, who were portrayed in full color and great detail, the men were positioned much lower, and were depicted entirely in dull gray, with indistinct features.[42] While viewers of this poster obviously understood that smokeless chimneys were a result, not a cause, of economic crisis, the disappearance of smoke from factory districts became a powerful symbol during the 1930s of domestic misery and male emasculation. As a result of the hardships of the Great Depression, many workers and their families no doubt welcomed the factory smoke that marked the return of a healthy economy as the country mobilized for the Second World War.

Disorder

Throughout the nineteenth century periods of social conflict led to concerns about the connection between crime and poor visibility. In *The Moral and Physical Condition of the Working Classes Employed in the Cotton Manufacture in Manchester* (1832), the physician James Phillips Kay (1804–77) claimed that "thieves and desperadoes" frequently took refuge in a district "surrounded on every side by some of the largest factories of the town, whose chimneys vomit forth dense clouds of smoke which hang heavily over this insalubrious region."[43] Visiting Manchester three years later, Alexis de Tocqueville was similarly impressed by its thick smoke, which turned the sun into "a disc without rays" and transformed the city's streets into a "dark labyrinth."[44] Commenting on the power of such imagery, Yi-Fu Tuan argues that the maze-like quality of nineteenth-century European cities provoked anxiety among visitors, particularly if these cities were not brightly lit: "To local residents their own neighborhoods of winding streets, dead-end alleys, and courtyards might seem familiarly complex and intimate. To strangers, however, it was a bewildering and frightening place to stray into as the sun began to set and the shadows lengthened."[45] Many middle-class commentators—obsessed with a supposed connection between darkness and crime—saw air pollution not primarily as a public health problem but as a catalyst and cloak for social disorder.[46]

Half a century after Kay's book appeared, fears about the association between crime and polluted air resurfaced. In 1881 the jurist Sir William Frederick Pollock (1815–88) claimed that dense fog not only interfered with trade and employment, but also created "dangers to life and limb and property" in darkened streets and provided a cover for "plunder, either by stealth or violence."[47] Pollock's assertions about the relationship between atmospheric conditions and crime were apparently borne out on 8 February 1886, when a dense, smoky fog provided cover for "roughs" who occupied Trafalgar Square and subjected the fashionable West End to a wave of vandalism and looting.[48]

Anxiety about class fundamentally shaped the ways in which contemporaries understood coal smoke. In 1880 Rollo Russell wrote an influential tract that directly implicated air pollution in class conflict. Arguing that "the presence of an overshadowing cloud of smoke produces moral evils," he asserted that the "one thing for which, more than any other, the poor of London express envy of the rich is the power of going at any time 'to the country.'" The fresh air of the countryside, he argued, ensured that even its poorest inhabitants possessed a "wholesome buoyancy of mind" and the "appearance of contentment." In London, however, "smoke and bad air" contributed to the "pallor, discontent, and ill-health" of its poor.[49] The following year the *Fortnightly Review* declared that improving the living conditions of the lower classes was not only a prudent health measure; it was also "an insurance paid by the rich against revolution."[50] The *Lancet* expressed similar views in a prominent article on urban problems that appeared in the same year, predicting that if the government did not take steps to improve the living and working environment of its citizens, "we may have a yet ruder awakening from the dream that a nation exists only for its upper classes to live in luxury, while its lower ones die in misery."[51]

Artists were particularly sensitive to smoke. In 1879, foreshadowing later debates over postimpressionism, the *Times* complained that smoke dangerously distorted ordinary perceptions of reality. A polluted atmosphere, it argued, caused artists to see things "not as they are, but as the painter thinks they ought to be."[52] Beneath the literal meaning of this statement lies the suggestion that pollution may inspire a revolutionary rejection of modern industry and the economic structures associated with it. The anxieties expressed in the *Times* article were not unfounded. William Morris's philosophy combined a critique of environmental pollution with an attack on capitalism. Speaking in Leicester in 1884, he declared that the continued transformation of the atmosphere into "dirt" could be halted only by overturning "the present gospel of capital."[53]

Shortly after he organized the Coal Smoke Abatement Society in 1898, Sir William Blake Richmond exclaimed, "What in the name of God's earth ... is the use of all our fine sculptures, our pictures, our embroideries, our buildings, if, a few weeks or so after they have passed from our hands, they are covered with coats of black and greasy dirt—grease that you can't

Figure 5.3. "The poor fogged-out painters in oil and water colours parading in the streets of Kensington, and singing in chorus." From *Punch*, 21 Feb. 1880, 78.

eradicate?"[54] Although such concerns appealed to many whose support Richmond sought to cultivate, they alienated others. Less than a year after the new group had formed, the *Builder* observed, "Business men often say that the antismoke movement is an amusement of faddists and try to brand it as a mere æsthetic craze. If there are two things which the average Englishman detests, they are the fad and the æsthete."[55] Although Richmond was a prominent member of the Royal Academy of Arts, he insisted that his antismoke activity was motivated by altruistic concerns for the nation's economic and physical welfare: "I am fighting for LIGHT, blessed light, not for myself alone, but for the people."[56] Despite such disclaimers, aesthetic considerations were important to many opponents of smoke. The president of the Royal Academy, Sir Frederick Leighton, was a prominent supporter of the National Smoke Abatement Institution in the 1880s. At an 1882 dinner marking the conclusion of the group's smoke abatement exhibition Leighton asserted that artists, because of their sensitivity to nature, were particularly averse to polluted air.

> Many a brother painter must regret with him the interminable hours, days, and weeks of enforced idleness spent in the continuous contemplation of the ubiquitous yellow fog, depressing the spirits all the more for recalling the memories of distant lands where the sun shone in the sky and shed its golden lustre over all things; where the fragrance of a thousand blossoms, not the soot of a thousand chimneys, was wafted in through open windows, and where grime did not blot out the heavenly face of nature.[57]

Richmond shared Leighton's opinion that smoke and fog interfered with the ability to see nature the way it really was, a defect that he associated with impressionism in painting.[58] He embraced romanticism in art and architecture, particularly that practiced by the pre-Raphaelite brotherhood.[59] Significantly in this regard, one of the members of the Coal Smoke Abatement Society was the pre-Raphaelite painter William Holman Hunt.[60]

While many artists decried fog and smoke, others—including J. M. W. Turner and Claude Monet—celebrated their visual effects.[61] Monet noted that London's fog came in many colors: "There are black, brown, yellow, green, purple fogs, and the interest in painting is to get the objects as seen through all these fogs." Monet once went so far as to claim that "without the fog London would not be a beautiful city."[62] The writer M. H. Dziewicki asserted that fog was a powerful catalyst to the imagination. For him fog was "terribly picturesque," for it suggested sublime scenes of "terrible calamity," death, and destruction.[63]

Loss

During the nineteenth century the twin forces of industrialization and urbanization distanced the past from the present by increasing the pace of life, by introducing new technologies and patterns of work, and by transforming the environments in which people lived. Reflecting on the enormity of these changes, the cultural critic Frederic Harrison (1831–1923) wrote in 1882, "The last hundred years have seen in England the most sudden change in our material life that is perhaps recorded in history."[64] This sentiment was widespread in the late nineteenth century, as Victorians sought to adjust to the transformations through which they had lived.

One of the most influential observers of the changes that shaped nineteenth-century Britain was the artist, critic, and reformer John Ruskin (1819–1900). In *The Seven Lamps of Architecture* (1849), Ruskin argued that modern innovations were artificial and devoid of humanity. The British people, he suggested, were losing their bearings, for "the place of both the past and the future is too much usurped by the restless and discontented present. The very quietness of nature is gradually withdrawn from us."[65]

Ruskin regarded the atmosphere in metaphysical terms, as symbolic of the relationship between society and the natural world. As an artist and a lover of landscapes, Ruskin was an avid observer of clouds and other atmospheric conditions. In the early 1870s he began to notice a change in the air. In a pair of lectures that he delivered before the London Institution in 1884, Ruskin argued that a "plague wind" had descended on Britain, defiling its atmosphere and changing the character of its clouds. Ruskin remained vague about the cause of this phenomenon, and some members of the assembled audience reportedly treated his words derisively, viewing them as proof that the eccentric sage was once more succumbing to mental illness.[66]

Commentators have tended to interpret these lectures as cultural critique, hardly pausing to consider whether the phenomena that Ruskin described in 1884 were real—the result of coal combustion and the eruption of Krakatoa the previous year. To Ruskin, the observation that the air had deteriorated was *both* a statement of fact and a sign of humanity's failings. In his view, people had desecrated nature, and people had a moral impera-

Figure 5.4. "The Clyde—Beauties of Scotch Scenery as Seen by Our Artist." Although Scotland was famed for its lochs and highlands, Linley Sambourne here emphasized its industrial character along the River Clyde in Glasgow. From *Punch*, 6 Aug. 1881, 51.

tive to undo the damage that they had caused. Ruskin's title for the lectures, "The Storm Cloud of the Nineteenth Century," made clear his view that the problem was linked to modernity.[67] Many shared Ruskin's view. Glasgow's medical officer of health maintained that his city possessed "a landscape which, as it left the hands of nature, was as fair as eye could see, but which now, when visible at all, discloses a bleared and blackened visage through a veil of smoke."[68]

Ruskin's antimodernism similarly found expression in his ideas for the Guild of St. George, a rural commune based on cottage production and the avoidance of steam power. Reverence for nature was central to the guild's principles; one of the articles in the vow to be taken by new members stated, "I will not kill nor hurt any living creature needlessly, nor destroy any beautiful thing, but will strive to save and comfort all gentle life and guard and perfect all natural beauty upon the earth."[69]

To some people, even the passage of time seemed to have accelerated. Expressing this view in 1882, the legal scholar and Liberal MP James Bryce (1838–1922) noted, "We live in a time when the past is vanishing with unexampled rapidity."[70] Bryce, who chaired the Commons Preservation Society, later served as a founding member of the National Trust for Places of Historic Interest or Natural Beauty.[71] A perception that the present was becoming increasingly disconnected from the past heightened people's interest in preserving tangible fragments of earlier times. Commenting on this characteristic of nineteenth-century culture, David Lowenthal argues that "no people since the Renaissance" had exhibited as much "antiquarian retrospection."[72] One way in which people expressed an interest in the past was by working to preserve its physical remnants, such as old buildings and monuments. In Ruskin's view, historic architecture "half constitutes the identity, as it constitutes the sympathy, of nations."[73] Using words similar to Ruskin's, the historian Charles Dellheim observes that "in an age of radical social and environmental disruptions the preservation of familiar landmarks helped satisfy deep-seated needs for stability and orientation."[74]

Such landmarks were threatened in many places by three forces of modernity. One of these was the demolition of old buildings and other artifacts to clear land for new buildings, streets, and railroad terminals.[75] To prevent this from happening, the biologist and Liberal MP Sir John Lubbock (1834–1913) introduced a bill in 1873 that would have created a commission with the power to prohibit the destruction of ancient monuments on private property.[76]

A second threat came from the overzealous "restoration" of churches and cathedrals, which William Morris and others blamed for falsifying and destroying England's architectural heritage. In 1877 Morris founded the Society for the Protection of Ancient Buildings, which pushed for government action. The ensuing debate over historic preservation prefigured later controversies over pollution, for both involved questions about the legitimacy of the state telling people what they could do on private property.

Coal smoke constituted a third way in which modernity jeopardized historic buildings across Britain. One historic site to be threatened by smoke during the late nineteenth century was Kirkstall Abbey, located in the northern industrial city of Leeds. Famous for its production of woolen cloth and garments, Leeds was also a center for boot making, iron and steel manufacturing, and, by the early twentieth century, automobile production.[77] As a result of industrial and household coal consumption, smoke, soot, and tiny particles of tar filled the air of Leeds, adhering to every surface and turning everything black. Although largely destroyed during the Reformation, the abbey's ruins remained an important landmark for those who sought to retain aspects of the city's preindustrial heritage. When the abbey and its adjoining property went up for sale during the 1880s, many feared that the ruins might be obliterated to make way for de-

Figure 5.5. Industrial smokestacks and household chimneys near Kirkstall Road in Leeds, ca. 1896. From Julius B. Cohen, *The Character and Extent of Air Pollution in Leeds* (Leeds, 1896), 4.

velopment. Preservationists breathed an enormous sigh of relief when the industrialist who purchased the site donated it to the city with the stipulation that it be preserved in perpetuity.[78] Although this action saved the abbey from immediate destruction, it did nothing to halt its deterioration from the effects of coal combustion.

Besides smudging buildings, coal smoke contained large quantities of sulphur dioxide. When it combined with water to form acid rain, it caused stone and mortar to flake and crumble. The air of some industrial cities was so acidic, noted Robert Angus Smith in 1859, that blue litmus paper turned red after only ten minutes of exposure to it.[79] Buildings across Britain faced similar disfigurement and deterioration. In 1800 Glasgow's numerous sandstone buildings had been a golden color; by the middle of the nineteenth century they were encrusted in black soot.[80] Describing Sheffield in 1861, the *Builder* observed that the air over the town was filled with "a thick pulverous haze . . . which the sun even in the dog days is unable to penetrate."[81] Conditions remained bad in Sheffield. Its town hall, an architecturally striking building completed at great expense in 1897, soon became "a mountain of soot without grace or distinction," in the words of one opponent of smoke.[82]

The present Palace of Westminster (more commonly known as the Houses of Parliament) was built more recently than many people realize. Although its architecture may suggest that of the late Middle Ages, the present building was constructed in the middle of the nineteenth century (and extensively repaired after suffering bomb damage in the Second World War). Huge sums were expended on its construction, which lasted for over a dozen years and required immense quantities of quarried stone. To the dismay of many the stone facade of the building began to exhibit signs of decay even before the building was completed. A special committee, which included prominent architects, engineers, geologists, and chemists, was appointed to discover the causes of the problem and to recommend solutions. The committee's report in 1861 attributed part of the decay to "the corrosive influence" of London's air, which contained, among other things, "carbonic acid, nitric acid, . . . several acids of sulphur, and occasionally hydrochloric acid."[83]

Instead of trying to reduce the severity of coal smoke, the chemist Augustus Voelcker searched for ways to lessen its impact. More research was needed, he asserted in 1864, "to avoid the employment of building materials which, although they stand the influence of the weather very well in the open country . . . , cannot be used with safety in some localities." Many of the building stones currently in use were not "capable of withstanding the pernicious chemical influence of a . . . town atmosphere."[84] By the early 1880s, the cost of repairing the damage that polluted air caused to the building was £2,500 per year.[85]

The Houses of Parliament were not the only structure affected by the acidic atmosphere of London. As one observer pointed out in the 1870s, most of the stones that adorned the exteriors of Westminster Abbey and Lambeth Palace were replacements for originals that had fallen victim to London's acidic atmosphere.[86]

One of the most unfortunate cases of deteriorating stone was that of the ancient Egyptian obelisk known as Cleopatra's Needle. This massive granite monument arrived in London in January 1878 after a journey that lasted more than four months and cost the lives of six sailors. After much discussion about where the nearly two-hundred-ton obelisk should be placed, it was finally raised into position on the newly completed Thames Embankment not far from Whitehall. The transfer of the obelisk from Egypt to Britain contributed to efforts to turn London into the world's greatest imperial capital. The acquisition and display of artifacts such as the Elgin Marbles and Cleopatra's Needle created powerful links between

Figure 5.6. The Tower of London was one of many historic buildings damaged by polluted air. In this 1920 photograph, the old stones in the upper portion of the arch form a sharp contrast to the newly replaced stones below. From Ministry of Health, *Interim Report of the Committee on Smoke and Noxious Vapours Abatement* [Cmd. 755], H.C. 1920, facing 3.

nineteenth-century Britain and the great civilizations of antiquity, and helped to suggest that British civilization was as significant historically as those of ancient Greece and Egypt. Expressing the hope that British civilization would last as long as had the great civilizations of the past, a letter to the *Builder* in 1878 suggested that Cleopatra's Needle would watch over London "for long centuries . . . , till, in fact, time's tooth gnaws through the granite itself." Only a few months later, however, experts warned that the obelisk was in jeopardy; the acidic atmosphere of London was causing the stone to deteriorate at an alarming rate. Sir Joseph Bazalgette (1819–91), the engineer who directed the construction of the Thames Embankment, urged the Metropolitan Board of Works to act quickly to prevent further decay and suggested applying a protective solution to its surface. Despite attempts to arrest the corrosive effects of London air on the obelisk, its stone continued to disintegrate. As one observer noted in 1890, "Cleopatra's Needle, which has endured unchanged for scores of centuries on the banks of the Nile, is already hastening to decay in the murky fogs of the Thames."[87]

Ancient relics gained much of their symbolic power from their proven ability to survive the ravages of time. In Ruskin's words, such objects transcended historical change and stood in "quiet contrast with the transitional character of all things . . . through the lapse of seasons and times and the decline and birth of dynasties."[88] Professor Sidney Colvin, a founding member of the Society for the Protection of Ancient Buildings, similarly maintained that a venerable monument or building evoked a sense of permanence and significance that modernity lacked, precisely because of "all the storms it has weathered and all the vicissitudes it has survived."[89] Threats to historic symbols—whether by demolition, flawed restoration, or atmospheric pollution—strengthened the sense that time was speeding up and that civilization had lost its moorings. If nineteenth-century civilization could obliterate buildings and monuments that had survived through the ages, many wondered whether historic social and political structures would fare any better. By an interesting coincidence, the same issue of the magazine carried another article on the question of historical permanence, entitled "The Integrity of the British Empire." Memories of the past formed an increasingly important part of British national identity as optimism about its future began to fade in the late nineteenth century. To these factors should be added concerns about the future of the empire. These anxieties assumed an increasingly nationalistic and racist tone in the late nineteenth and early twentieth centuries.

Nature and Social Stability

While many found transcendence in historic buildings, others looked to the natural world. In 1910 the environmental activist and sex radical Edward Carpenter (1844–1929) argued that coal smoke was a spiritual offense, and not only because it damaged the facades of churches and chapels. "It is sad," he declared, "that all our appreciation of holiness should be piled on these things to the neglect of the beauty and sacredness of human life, and that while we sing our hymns to God on the Sundays, all the week we should empty our slops into the streams, and pour foul volumes of smoke into the glorious and crystal vault of heaven. I am sure that God does not like that sort of worship!"[90] The social reformer Octavia Hill (1838–1912) would certainly have agreed. Despite being a granddaughter of the nature-phobic Thomas Southwood Smith, Hill had much more in common with the romanticism of her mentor John Ruskin, who bankrolled her early experiments in housing reform. In an 1877 article in the *Fortnightly Review*, she claimed that nature transcended social and chronological divides, constituting

> a link between the many and through the ages, binding with holy, happy recollections those who together have entered into the joys its beauty gives—men and women of different natures, different histories, and different anticipations—into one solemn, joyful fellowship, which neither time nor outward change can destroy—as any people bound together by any noble common memory, or common cause, or common hope.[91]

This belief that nature and history constituted a source of social stability was later embodied in the National Trust, a group that Hill was instrumental in founding. As we shall see, the National Trust provided important support to the smoke abatement movement, and many clean air activists considered their fight against smoke as an important contribution to the preservation of historic sites and "nature."

Even before smoke came to be seen as unhealthy, it was recognized as a source of grime. As John Simon noted in 1848, smoky city air led many people to "keep their windows shut, breathing a fusty and unwholesome air in the hope of excluding the inconvenience."[92] Many worried smoke and soot not only discouraged ventilation, but also undermined domestic order by transforming conscientious and hearty immigrants from rural areas into physically and morally degenerate slum dwellers. According to Rollo Russell, it crushed "attempts at cleanliness and neatness among even the most scrupulous of the poor. Wives of labouring men ... find the task of keeping their houses clean too hard for them and give it up in despair. A forced neglect thus eats into the domestic happiness and disheartens the spirits of the best of them."[93]

Middle-class observers in the late nineteenth century often maintained that exposure to nature was vital for social cohesion. Long before the aerial bombing of the Second World War forced the evacuation of children from London, philanthropic groups in many of Britain's largest cities began to organize "country holidays" for urban children. Organizers frequently stressed not only the health benefits of such outings, but also their "civilizing" effects. In 1877 Samuel Barnett (1844–1913), an Oxford-educated vicar involved in the settlement-house movement in the East End of London, and his wife Henrietta Barnett (1851–1936) began what later became known as the Children's Country Holiday Fund to make this possible. By the time of the First World War, the fund had provided country holidays to nearly one million children.[94]

Some argued that no corner of Britain offered a sanctuary from the problems of city life. To William Morris, London was a "spreading sore" that was destroying "field and wood and heath without mercy and without hope, mocking our feeble efforts to deal even with its minor evils of smoke-laden sky and befouled river."[95] Despite the radical nature of Morris's critique, many shared his concerns. The *Lancet*, one of Britain's leading medical journals, described London in 1881 as an "ever-increasing Babylon, which, like a malignant growth, is rapidly eating its way into the verdant country and disfiguring the face of nature with jerry-built villas, each of which vomits forth its contribution to the general pall of smoke."[96] Elaborating on this theme, the plant expert Robert Holland asserted that "the air of the whole of England" was polluted. Speaking in Manchester in the late 1880s, he claimed that the "grand old ancestral trees, which once flourished throughout the length and breadth of the land and added so greatly to the beauty of an English landscape, are gradually dying and rotting as the foul, polluted air reaches them."[97] Writing in 1907, the principal of Dalton College at the University of Manchester complained that, because of smoke, "the moors in Derbyshire are dirty to sit down upon, and the sheep upon them are dirty in colour. Even the Lake District is not entirely free."[98] This reference to the Lake District would have had special significance for many readers, as the area was associated not only with

Figure 5.7. "A Dream of Green Fields," by J. Bernard Partridge. Philanthropic efforts aimed at providing poor urban children with a holiday in the countryside emphasized the importance of even a temporary respite from coal smoke. From *Punch,* 10 Aug. 1904, 93. Reproduced with permission of Punch Ltd.

Wordsworth but also with Ruskin and some of the earliest battles to protect the countryside from development.

Criticism of urban industrial society was not limited to a particular political ideology. Socialists, liberals, and conservatives alike expressed a strong

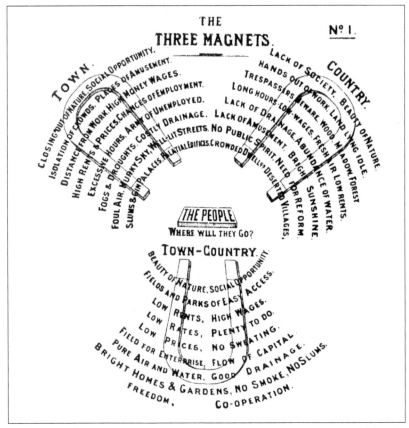

Figure 5.8. "The Three Magnets," Ebenezer Howard, *To-morrow: A Peaceful Path to Real Reform* (1898), facing 8.

interest in nature and used environmental arguments to criticize the present and to provide a vision of the future that they hoped to create.[99] As early as 1843 a journeyman bookbinder in Manchester, Benjamin Stott, remarked that he had "had few opportunities of looking on the sublime and varied face of nature, save through an atmosphere polluted with smoke belched from a thousand chimneys in this large manufacturing district."[100] Half a century later Ebenezer Howard (1850–1928), the promoter of garden cities, used his famous "Three Magnets" diagram to explain that his ideal community, "Town-Country," would be free from the "fogs," "smoke," "foul air," and "murky skies" found in urban places.[101]

The social and political turmoil of the late nineteenth century gave air pollution a cultural resonance that it had never before possessed. At a time of economic depression, growing foreign competition, and increasing class conflict, smoke and fog bolstered already existing fears that the future of Britain's urban and industrial society was in peril. Contemporaries began to imagine smoke not as a sign of industrial productivity but as an augury of social disorder, environmental imbalance, economic decay, and biological degeneration.

CHAPTER 6

Degeneration and Eugenics

A person who lives in the open air or inhabits a well-ventilated and carefully-appointed house on the slope of a hill has a very different appearance from the ordinary town dweller or the factory lad. The one is florid-looking, with hard, well-developed muscles and great strength; the other is pale-faced, has flaccid muscles, and is comparatively weak as regards physical power.
—*Alfred Carpenter, 1877*[1]

H. G. WELLS'S SCIENCE fiction masterpiece *The Time Machine*, published in 1895, centers on the idea of degeneration. The protagonist of the novel, having traveled thousands of years into the future, discovers that humanity as we know it no longer exists. Instead, it has evolved into two distinct species, whose struggle against each other mirrors the class conflict of nineteenth-century Britain. The descendants of the Victorian middle and upper classes, he learns, have atrophied mentally and physically to become weak creatures known as Eloi. The descendants of the working class, in contrast, have turned into the strong and predatory Morlocks, who use the Eloi as food.[2] For many readers of *The Time Machine* in the 1890s, the prospect of the rich and the poor becoming biologically distinct—and increasingly antagonistic—was not confined to science fiction. Inverting the earlier notion that disorder in the natural world ("dismal swamps of vast extent, reeking with masses of decaying vegetable and animal matter") was polluting the air and harming human beings, many now feared that humanity's use of technology was polluting the air with smoke and harming nature.[3] As the lord provost, that is, mayor, of Glasgow put it in 1890, because of smoke, "Trees die, flowers will not grow, even our grass degenerates into weeds, for obnoxious forms of vegetation can resist unwholesome conditions."[4] Influenced by the scientific thinking of the era, many believed that the conditions of urban and industrial society were causing human beings to degenerate as well.[5]

Although the theory of degeneration is often associated with eugenics, this linkage is only half of the story. While eugenicists blamed degeneration on faulty heredity and looked to selective breeding as the solution, environmental reformers attributed it to unwholesome surroundings— particularly those found in crowded urban areas. As social Darwinism and eugenics gained strength during the late nineteenth century, antismoke activists articulated a fundamentally different explanation, and cure, for degeneration.

It was widely accepted that the healthiest people in Britain's cities were recent arrivals from rural areas of England, Scotland, or Wales—but not Ireland. Reflecting a deep-seated prejudice against Irish immigrants, many commentators assumed that they had already degenerated under the debilitating influences of Roman Catholicism, poverty, ignorance, and exploitative landlords. One writer asserted in 1866 that most of those who migrated from Ireland to England "constitute but sorry specimens of the Irish peasant."[6] Many viewed urbanization as the chief cause of the problem. As the physician John Milner Fothergill put it in 1889, "Town life is seen to have a malignant and sinister effect upon the physique." Fothergill described the inhabitants of cities as "a doomed race" whose descendants would scarcely survive beyond a few generations without "infusions of new blood" from rural areas where people benefited from "plenty of fresh air."[7]

These ideas were disseminated widely on both sides of the Atlantic in the early years of the twentieth century. One prominent popularizer was Madison Grant, who served as a trustee of the American Museum of Natural History and was the chair of the New York Zoological Society. His 1916 book, *The Passing of the Great Race, or the Racial Basis of European History,* is an exemplar of early twentieth-century scientific racism. Displaying a level of prejudice that was widespread both in Britain and the United States, Grant asserted that northern Europeans were racially superior to all other ethnic groups. Disconcertingly for him, however, this supposed hereditary superiority did not appear to function in all environments. Echoing the dystopian predictions of *The Time Machine,* Grant argued that "heavy, healthful work in the fields of northern Europe en- ables the Nordic type to thrive, but the cramped factory and crowded city quickly weeds him out."[8] Cities were unnatural and dangerous, he im- plied, because they appeared to turn the Darwinian law of the "survival of the fittest" against the group that he assumed was inherently superior.

Historians who have discussed the idea of degeneration have tended to focus on eugenics and have largely ignored the ways in which the idea shaped the efforts of environmental reformers.[9] Public health reformers and eugenicists alike believed that they, as experts, possessed the right to act in ways that would counteract the threats to health and civilization, but this similarity disguised a fundamental disagreement about the ame- liorative potential of environmental improvements. The primary differ- ence between eugenicists and environmental reformers was not a dis- agreement about whether poor people were generally shorter and weaker than those who had more money, but whether this was the result of heredity or environment.

One of the leading advocates of the view that class distinction was rooted in biology was the physician Robert Jones. Addressing the Society of Arts in 1904, Jones declared that he was "convinced . . . of the existence of a deterioration among the lower classes and their issue." In his esti- mation, the ideal human type was "the thoughtful, well-balanced brain worker." At the other end of the spectrum lay "extreme degenerates," who formed the antithesis of the bourgeois ideal. They included "the physically impaired and the mentally unstable." According to Jones, poor people pos- sessed a "hereditary predisposition to break down under the depressing effects of town life." At the same time, however, Jones claimed that "too much is made of heredity and too little of environment." Efforts to send poor children from "airless and sunless London" to spend time in the countryside showed that "much is done by a new environment to contend against faulty heredity." If the working classes could not be dispersed along with industry into the country along the lines of Ebenezer Howard's garden cities, the next best thing would be to bring a bit of the country into the city, in the form of playgrounds and "physical training . . . in the open air."[10]

The Air of Towns

Many commentators believed that the essential difference between healthy rural areas and unhealthy cities lay in the different types of air found in each place. In an alarmingly titled book, *The Danger of Deterioration of*

Race from the Too Rapid Increase of Great Cities (1866), the Manchester physician John Edward Morgan (ca. 1829–1892) summed up this difference. Echoing Engels, Morgan contrasted the countryside, where workers spent their days "under an open sky, while the lungs are filled with fresh air," with the city, in which many labored in "the vitiated atmosphere of confined workshops" and lived in narrow, poorly ventilated streets. Morgan argued that to prevent "race deterioration," cities had to be made healthy by improving their air: "Cellar dwellings should be closed up, courts and alleys cleared out, and the sites which they occupied left open to serve the double purpose of air-shafts and play-grounds; streets also should be widened, factories and workshops, instead of being piled up in the centre of our towns, should be scattered over the country in airy and healthy localities."[11]

Statements about the unhealthy nature of urban environments often contained explicit or implicit claims about the biological inferiority of the lower classes. In an address to the National Association for the Promotion of Social Science in 1874, the chemist and Liberal MP Lyon Playfair (1818–98) argued that "crowded and unhealthy" environments led inevitably to "low morals and low intelligence." Making his degenerationist argument perfectly clear, he asserted that in such places "the condition of human beings is scarcely above that of animals—where appetite and instinct occupy the place of the higher feelings."[12] Although Playfair's comments were ostensibly about environmental conditions, the social nature of his derisive characterizations of urban residents is clear.

Arguments that sought to justify economic inequality in terms of ethical or physiological differences gained added saliency in times of intensified class conflict. Degeneration theory first gained prominence during the Chartist agitation of the 1840s,[13] faded during the mid-nineteenth century, and reemerged in its closing decades as workers sought greater economic and political power. The outburst of working-class political and trade union activity prompted those in positions of authority to think once again about social inequality, of which health was a major component. In 1878 the Anglican bishop of Peterborough, William Connor Magee, warned that class divisions were "widening and deepening day by day, as the envious selfishness of poverty rises up in natural reaction against the ostentatious selfishness of wealth." If nothing were done to ameliorate the suffering of the poor, he warned, their desire for a better life might eventually call forth "an exceedingly great army more to be dreaded than the invading hosts of any foreign foe."[14]

Imperialism and the Body Politic

Concerns about the biological health of the urban working class contributed to growing anxiety about the economic and imperial fitness of
Britain in the late nineteenth century. In 1887 Lord Brabazon warned that
if present trends continued, this "would ultimately lead to a degeneration
of the race and to national effacement."[15] Subsequent reports seemed to
bear out these predictions. Over half of military recruits in 1887 and 1900
were shorter than five feet six inches tall, compared to only a tenth of
the recruits in 1845. Despite assertions that this indicated universal decline
among urban residents, recruits hardly constituted a representative sample.
Enlistment was voluntary, and many men with other employment options
chose not to enlist, especially when higher-paying jobs became available
elsewhere.[16] Fears of degeneration peaked during the second Anglo-Boer
War (1899–1902). Instead of achieving a quick victory against what jingo-
istic British newspapers had portrayed as a "handful of Dutch peasants,"
the war was long and full of setbacks.[17] The news that over three-fourths
of the men volunteering for military service in Manchester had been turned
away as unfit prompted many commentators to blame Britain's military
difficulties on the poor health and slight stature of its soldiers.[18]

These allegations generated enormous political pressure on the government to do something. It responded in 1903 by appointing the awk-
wardly named Inter-Departmental Committee on Physical Deterioration.
Many of the witnesses who testified before the committee attributed the
unhealthiness of cities to environmental factors, particularly a lack of sunlight and fresh air. In contrast to later environmentalists, however, most
of those who addressed the committee limited their attention to the indirect health effects of smoky air. Believing that germs were the sole direct
cause of disease, they asserted that smoke harmed health by interfering
with the ability of nature to prevent and heal illness. They pointed out that
smoke reduced the amount of sunlight and ozone—both widely recognized as disinfectants—that could penetrate cities and destroy their germs.
In addition, they noted that smoke and soot made it difficult for oxygen-
producing plants and trees to survive in urban parks. D. J. Cunningham,
an Edinburgh anatomy professor, testified that "the more nearly you can
approach the rural life, the greater amount of certainty you will have that
there will be an improvement in the physical conditions of people." Another witness argued that the leading cause of ill health in cities was "the
condition of the air. There is very little sunlight—only some two-fifths of
what there might be under other conditions, and there are a great many

deleterious elements in the air."[19] The Sheffield smoke inspector William Nicholson made the same point, asserting in 1905 that "a proportion of the physical degeneracy of the human race . . . is directly traceable to the constant inspiration of impure air."[20] The report of the Inter-Departmental Committee on Physical Deterioration concluded that health had actually improved on average during the preceding fifty years. Rejecting hereditary explanations, the committee held that disparities in the health of various places were largely caused by poor enforcement of public health measures.[21] It recommended that greater attention be given "to the preservation of open spaces with an abundance of light and air," and it urged stricter legislation and better enforcement to reduce "pollution of the air by smoke and noxious vapours." Reflecting the growing view that domestic smoke contributed substantially to the polluted air of cities, it added that efforts should be undertaken to convince ordinary householders of the importance of smoke abatement.[22]

The Struggle over Eugenics

In sharp contrast to the committee, which searched for environmental causes of ill health, eugenicists were convinced that degeneration was hereditary and was impossible to ameliorate by improving the environment of Britain's cities. Eugenicists maintained that environmental reforms were not merely futile, but actually counterproductive.[23] The London surgeon Hugh Percy Dunn, for example, argued in 1904 that many people who formerly would have died were now surviving because of improved public health measures. Implying that a clear distinction existed between the fit and the unfit, he asserted that as a result of

> hygienic science . . . the healthy community is leavened, so to speak, with individuals whose physical degeneracy has not been acquired, but inherited, and whose physical weakness naturally tends numerically to become more and more manifest . . . as our health laws become more and more perfect. Surely the greater the care exercised in the preservation of physically degenerate individuals the greater must be the increase in the degenerate examples of our race. In olden days presumably every man and woman who reached adult life was an illustration of the law of the survival of the fittest; in these days the presumption is that such a law has ceased to be operative, and that the weak and the strong grow up together.[24]

The statistician and eugenics crusader Karl Pearson (1857–1936) similarly argued that improvements in public health had allowed people to survive who should have been eliminated through natural selection. Pearson also warned that the least healthy sixth of the population bore children at such a prolific rate that it was producing half of the children born each year. He insisted that if Britain were to maintain its global power, it had to be a "homogeneous whole" rather than "a mixture of superior and inferior races."[25] Ethel Elderton, a researcher at the Galton eugenics laboratory, shared Pearson's desire to spread this message to a popular audience. In a 1909 book titled *The Relative Strength of Nurture and Nature*, Elderton asserted, "The views of philanthropists and of those who insist that the race can be substantially bettered by changed environment appeal to our sympathies, but these reformers have yet to prove their creed. So far as our investigations have gone they show that improvement in social conditions will not compensate for a bad hereditary influence."[26]

Poor people, argued social Darwinists and eugenicists, were poor because they were less physically and mentally capable than their social superiors. According to this line of thinking, the environment exerted an insignificant influence on health because "inferiority" was largely inherited. As Gerry Kearns has observed, "the notion that social characteristics were carried in the blood gave rise to the idea that the differences between the classes could be passed, biologically, from one generation to the next, almost literally making the poor a race apart."[27] Expressing such a sentiment, Dr. John Milner Fothergill declared in 1889 that residents of cities displayed "a reversion to an earlier and lowlier ethnic form. While the rustic remains an Anglo-Dane, his cousin in London is smaller and darker, showing a return to the Celto-Iberian race."[28] Fothergill implied that as a result of miscegenation, urban members of the English "race" were slowly coming to resemble less "advanced" people in appearance.[29] The Fabian intellectual Sidney Webb (1859–1947) shared some of Fothergill's assumptions, arguing in 1907 that if the existing high rate of reproduction among the racially "unfit" continued it would lead to either "national deterioration, or . . . this country gradually falling to the Irish or the Jews."[30] By the time of the First World War, critiques of urban life founded on ethnic and class prejudice were sufficiently influential to shape the ways that people in other countries viewed Britain. In 1916, for example, Madison Grant warned that "if England has deteriorated . . . it is due to the lowering proportion of the Nordic blood and the transfer of political power from the vigorous Nordic aristocracy and middle classes to the radical and labor elements, both largely recruited from the Mediterranean type."[31]

Long before civil rights advocates and environmentalists came together in the 1980s to decry environmental racism and call for environmental justice, those in power used their privileged positions to shift the burden of pollution to others. Recognizing that poor and working-class people were often exposed to higher levels of pollution than those experienced by other sections of the population, some commentators sought to justify this through hereditary arguments. In their view, although high levels of pollution would injure the delicate sensibilities of the elite, they would cause no harm to the poor, who were already debased. A sanitary inspector from London observed in 1899 that there was less acceptance of smoke in upper-class areas than elsewhere: "The character of a neighbourhood was a determining factor. Thus, in Westminster ... a much less amount of smoke would be considered a nuisance than would be tolerated in Bethnal Green, where the factories are surrounded by the houses of the people who work in them."[32] A few years later, in 1905, the London physician Thomas Glover Lyon expanded on the notion that workers, being less refined than their social superiors, could be exposed to higher levels of pollution without detriment. "Adults engaged in manual labour," he asserted, "do not require pure air such as is found in the country." Yet "people engaged in mental work, and those whose occupation preclude [*sic*] a sufficient amount of exercise, fail in health unless from time to time they make excursions into the country."[33] The Birmingham physiologist and eugenicist David Fraser-Harris (1867–1937) similarly maintained that the lower classes did not deserve the same environmental conditions as did the middle classes, because the former lacked the ability to distinguish between good and bad air.[34]

In addition to using eugenics to rationalize complacency in the face of stark examples of social and environmental inequity, some eugenicists called for more decisive measures. In 1881 the *London Quarterly Review* declared that degeneration was the result of "not lopping aged stems, pruning weakly foliage, trimming budding growths, but sapping the strength of the most glorious trees, and leaving them to rot from generation to generation."[35] Social Darwinists argued that if such steps were justified in perfecting the natural world, they ought to be applied to humanity as well. In 1909 David Fraser-Harris chillingly declared that it had "come to be a question in the minds of many how far we are justified in impeding the national march onward towards better things by fostering this large crowd of traitors and camp-followers in the national army."[36] This rhetoric implied that efforts to help the poor were counterproductive. According to proponents of eugenics, such a policy was dangerous; instead of allowing the weakest and least useful members of society to succumb to "natural" forces, it would help them to survive and multiply. Fraser-Harris left little doubt as to his own views on the matter. Viewing sick and weak human beings as little more than plants to be pruned, he suggested setting up a tribunal that would "be given the power of conducting those unhappy ones to an euthanasia by anæsthesia."[37]

Although few adopted this tone, many people found it persuasive. In 1912 eugenically motivated legislation was introduced in Parliament.[38] If passed, it would have placed special controls on individuals who were "incapable . . . (1) of competing on equal terms with their normal fellows; or (2), of managing themselves and their affairs with normal prudence."[39] The *Nation,* a Liberal magazine published in London, argued against the bill for three primary reasons. First, it followed earlier opponents of eugenics by arguing that heredity was not everything, observing that "the defects of the parent count for little if the conditions after birth can be equalised." Second, it argued that "talk of imprisoning the unfit and sterilising the degenerate" was "a partially conscious attempt to cripple social reform" on the part of the Conservative party. The real solution to the problem of "mental deficiency," it argued, would come from higher standards of living—something that the Conservatives opposed. Third, it suggested that the bill would grant the government enormous powers to infringe on personal liberty, particularly that of poor people. Individuals deemed incapable of managing their own affairs might be institutionalized and lose their power to "bargain for wages or change their masters." Even more chillingly, it argued that the bill amounted to "a first essay in the scientific breeding of the poor."[40] Decades before the Nazi rise to power in Germany, the question of what to do with the "unfit" was widely discussed in many countries. Yet in contrast to the situation elsewhere, including in the United States, sterilization never became official policy in Britain.[41]

Many believed that degeneration primarily affected poor and working-class individuals. Those who asserted that the lower levels of society were degenerating disagreed about what caused it and why it hit poor people more than wealthy ones. Some observers blamed the phenomenon on dark, ill-ventilated slums, far from nature. Middle- and upper-class individuals, by contrast, often lived in more "natural" surroundings, with parks and gardens nearby if in a city, and with meadows and fields if in suburbs or rural areas. Others argued that poor people were susceptible to degeneration because they were racially inferior to their social "betters." Referring to the working class, John Edward Morgan declared in 1866 that he had "been much struck by the singular want of stamina which characterises them." Foreshadowing later degenerationist rhetoric, he argued that few of the poor were "of that calibre from which we might expect either vigorous and healthy offspring, or arduous and sustained labour."[42] W. R. E. Coles, the honorary secretary of the smoke abatement exhibition held in London in 1881–82, suggested that the poor were particularly susceptible to "debasement" in a polluted urban environment. Speaking at a meeting of the Sanitary Institute, he maintained that a "darkened and polluted atmosphere" caused a decline in the "tastes and moral tone" of those surrounded by it.[43] According to this logic, the lower classes were unable to escape the influence of environmental conditions, but the elite were able to transcend the "laws of nature."

Commentators disagreed about whether degeneration would simply make the working class weak or actually transform it into something that was no longer truly human. In *The Time Machine,* Wells suggests that the consequences of degeneration are determined by social class. The novel portrays degeneration leading to physical atrophy among the middle class and turning the working class into brutish and amoral animals. Such views were by no means limited to eugenicists. In 1881 Lord Brabazon claimed that the "English race" was "pre-eminent in the achievement of feats of agility and strength." British explorers and athletes were scaling mountain peaks that "never before, since the foundation of the world, had been forced to acknowledge the supremacy of man." Expressing a clear social Darwinist conflation of biological and social hierarchy (and equating cities with the poor), Brabazon asked, "Do our athletes, our sportsmen, our travellers, our mountaineers, issue from the crowded lanes of overgrown cities? Are not their homes to be found rather amongst the pleasant places of the earth, in rural manor-houses, in retired parsonages, in country villas, or in the healthy portions of well-to-do towns, in the midst of comfort and plenty, with every means of exercising the healthy bodies which they have inherited from healthy and well-to-do progenitors?" Using the authority and rhetoric of science to legitimate his views of social hierarchy, Brabazon suggested that working-class aspirations had exceeded the parameters established by nature. "Nature is stern—she has no compassion—as men sow, in like manner shall they reap." The only way to avoid being destroyed was to obey the "laws . . . laid down by Providence for the guidance of man."[44] Repeating this theme two years later, he maintained that "the laws of nature . . . can never be evaded without punishment falling swiftly on the head of the evader."[45]

Brabazon's words were not simply allegorical. As Nancy Stepan notes, nineteenth-century Europeans used the concept of degeneration to provide scientific justification for "differences that were 'natural' between human groups" and to condemn those who transgressed social boundaries.[46] This ideology gained legitimacy by appearing scientific: injunctions about the necessity of obeying the "laws of nature" served to naturalize unequal social relations.[47] Daniel Pick similarly argues that elites, worried about the growing political power of the working class, hoped to reassert authority by focusing attention on the supposed physical deterioration of the "lower orders."[48]

British clean air activists of the late nineteenth and early twentieth centuries were among the most significant opponents of eugenics in Britain. Although many of them shared with eugenicists a belief that degeneration was occurring and that class had a biological dimension, they insisted that the bodies and morals of working people could be improved through environmental reform. Brabazon, who later helped to form the Coal Smoke Abatement Society, declared in 1884 that "deterioration in physique is in great measure due to three causes—viz., overcrowding, a want of adequate and properly cooked food, and lack of sufficient air and exercise."[49] Three years later, he reiterated this point, arguing that people were deluded if they thought that there was "something in British flesh and blood which is able to withstand the deteriorating influences of bad air and food, and want of healthy exercise."[50]

By arguing that the environment played a major role in shaping the health of individuals and groups, smoke abatement activists—along with environmental reformers who worked in other fields—challenged eugenicists' assumptions that poverty, ignorance, stunted growth, and high levels of disease and death were inherited traits. Though both environmentalists and eugenicists believed that the health, stature, and stamina of the population had declined in recent decades, they disagreed fundamentally about why this was so and what to do about it. Environmentalists attributed the worsening condition of the urban working class to rising levels of pollution, crowded living conditions, unhealthy work, and lack of exercise, all of which could be ameliorated by altering the social, economic, or technological status quo.

Eugenicists, however, believed that the problem was not inadequate help for the least fortunate members of society, but too much coddling. In their view, the achievements of which many Victorians were most proud—rising standards of living among the working class, the spread of water and sewer networks, and higher life expectancies—were actually costly mistakes. They alleged that public health "improvements," by making it possible for a larger proportion of the "unfit" to survive and reproduce, had the paradoxical effect of lowering the overall health and vigor of the nation.

These rival ways of seeing the world were reflected in the language that each side used to refer to the phenomenon that troubled them both. Eugenicists often used the term "racial degeneration," which implied that the change was irreversible and could be passed to all of one's descendants. Environmentalists preferred to call it "physical deterioration," which suggested that the problem was superficial, individual, and reversible.

During the early twentieth century eugenic ideas achieved great prominence throughout the world and led, in some places, to coercive policies such as forced sterilization of "degenerates."[51] That such draconian measures were resisted in Britain may have been due in no small part to a small but persuasive group of smoke abatement campaigners. It is to their work that we turn next.

Chapter 7

Environmental Activism

What a difficult and hard-working time we had. To fight the apathy and
negligence of six hundred years, to induce people to believe that the
state of the air was artificial and not natural, and that each and all
of us were individually in part responsible.
—*H. A. Des Voeux, 1936* [1]

DURING THE FINAL DECADES of the nineteenth century, a growing
number of reformers in Britain sought to protect and improve the environment of both cities and the countryside. Much of this activity took
place within existing voluntary associations. These groups, which expanded
rapidly during Victoria's reign, provided an important public forum for
women, middle-class professionals, Protestant nonconformists, and industrialists, many of whom resented the political power held by men,
aristocrats, Anglicans, and landowners.[2] One of the forces that spurred
the emergence of such groups was concern about the increasing difficulty
of observing and influencing the behavior of the lower classes. This concern was particularly prominent in voluntary societies involved in public
health and other aspects of social reform.

One of the most important of these groups was the National Association for the Promotion of Social Science, established in 1857. The Social
Science Association, as it was often known, sponsored a series of exhibitions throughout Britain during the 1870s to educate the public about the

"science of sanitation." The Ladies' National Association for the Diffusion of Sanitary Knowledge, which affiliated with the Social Science Association in 1859, worked throughout England to give tracts and lectures to poor urban residents. This organization also exerted a powerful influence over the development of "mothers' meetings," which up to a million women and children attended every week for lectures on parenting skills and sanitation. Given this level of popularity, it seems likely that such gatherings provided information that participants found useful. Beneath the stated goals of this activity some scholars detect a covert attempt at social control. The meetings not only provided an opportunity to reform the habits and values of workers and the poor but also to monitor them for signs of resistance to economic and political authority, such as mass protests, riots, and strikes.[3]

Middle-class commentators frequently considered poor people as another form of out-of-control nature. Social reformers often assumed that unhealthy environments produced poverty, ignorance, and crime. Convinced that poor people lacked the moral discipline and cleanliness that were central to "respectable" identity, reformers sought to regulate and discipline the bodies and behavior of the lower orders.[4] In 1880 the Birmingham Ladies' Association for Useful Work sponsored a lecture series on "the laws of health." Reporting on these lectures to the city health department, the honorary secretary of the association, Susan Martineau, conflated concerns about unhealthy air with concerns about class conflict. Environmental improvements and education, she argued, would raise the morals of the poor and promote class harmony. Such ideas pervaded the lecture series. After one of the lectures the speaker asked rhetorically, "May not the deterioration in the state of the blood caused by want of sufficient pure air . . . account for much of the immorality and crime common in unhealthy and overcrowded dwellings and neighbourhoods?"[5] Continuing the linkage between social class and intelligence, the report of the association for that year tellingly revealed that "many of our ladies prefer working amongst the *very* poor, but it is at least questionable whether they do not effect more practical good amongst those whose intelligence leads them to apply the knowledge they gain to . . . good purpose."[6]

The Ladies' Sanitary Association printed one and a half million copies of tracts on health between 1857 and 1881. Aimed at a working-class audience, they were cheap, used simple language, and were highly pedantic. The titles covered a diversity of topics, including nutrition, parenting skills, and the relationship between air quality and health. The latter topic inspired tracts such as "The Mischief of Bad Air," "The Worth of Fresh Air," "The Cheap Doctor: A Word about Fresh Air," and "What Can Window Gardens Do for Our Health?"[7] Although it is unclear how many working-class people actually read these half-penny tracts, their publication indicates that many middle-class reformers were deeply concerned about the air of large cities and its effects on the poor.

Smoke abatement groups formed in several large towns and cities in the late nineteenth century. The leading provincial antismoke society, whose founding preceded that of the London group by several years, was the Manchester and Salford Noxious Vapours Abatement Association. Although Manchester consumed substantially less coal than London, it did so within a more concentrated area. As a result, some claimed that its air was even more polluted than that of the metropolis.[8] The association, as its name suggests, attacked both coal smoke and chemical emissions, but it devoted most of its attention to smoke—both industrial and domestic. During the 1870s and 1880s it sponsored public lectures, classes, and publications that were intended to educate the public about the benefits of clean air and to promote gas cooking and heating appliances in place of coal fires. As a result of the group's efforts, the term "air pollution"—virtually unheard of before the 1870s—became a familiar part of the English language.[9]

Assigning Responsibility

Discussions of air pollution during the late nineteenth and early twentieth centuries frequently revolved around whether the factory or the household was the greater offender. No simple answer existed, for the proportion of smoke coming from factories and houses varied considerably between different cities. London in 1880 contained approximately six hundred thousand houses and six times as many fireplaces.[10] These houses, numerous as they were, used less than half of the coal consumed within the metropolis in the 1870s; most of the consumption took place in chemical works, potteries, sawmills, engineering works, coal-gasification plants, and other industrial facilities. Railroads and riverboats, which for statistical purposes were excluded from the industrial category, were also heavy users of coal in London.[11]

Although industry accounted for the bulk of London's coal consumption, most of its smoke came from private houses. How could this be? The answer lies in the fact that household stoves and fireplaces produced far more smoke from a ton of coal than did industrial furnaces and boilers.[12] According to a 1904 estimate, even during the summer months well over half of the smoke in London came from houses.[13]

Because of the dearth of reliable statistics for each place and the impossibility of knowing exactly how much smoke was produced from every fireplace, stove, and boiler, people often sought to infer the ratio of domestic to industrial smoke by comparing air quality on Sundays, when most factories were idle, with that on other days. If the air of a city was not markedly clearer on Sundays, they concluded that private houses produced most of the smoke. Rollo Russell, for example, noted in 1880 that "some of the very worst fogs have occurred on Sundays, and Christmas Day 1879 was nearly dark."[14] In contrast to London, where many people blamed household fires for causing the bulk of the smoke problem, observers often considered industry to be the main culprit in other large cities. At an 1890 meeting of the British Medical Association the medical officer of health Sidney Barwise argued that this comparison showed that most smoke, at least outside London, came from factories: "A glance at any manufacturing town on a Sunday and again on a working day proves at once that the smoke cloud does not come from the dwelling-houses."[15] Referring to Manchester, the *Lancet* asserted in 1894 that "manufacturing smoke, and that from small works especially, accounts for a large proportion of it—in summer for most—though the thousands of domestic fires contribute their share."[16] This view was soon embraced by the Committee for Testing Smoke-Preventing Appliances. Even though this group was supported by a number of factory owners, it conceded that industry produced most of the smoke that hung over manufacturing areas.[17] Unlike antismoke activists in London, those in Manchester sometimes argued that the combustion of coal in household fires was insignificant. The chemist Charles Estcourt, in an 1887 lecture sponsored by the Manchester and Salford Noxious Vapours Abatement Association, asserted that domestic coal fires produced less than "one-hundredth part of the sulphurous acid and black smuts that are evolved from the combustion of coal under large boilers."[18]

But comparisons of smoke between Sundays and workdays were subjective and anecdotal and led to conflicting claims. Others argued that the existence of smoke on Sundays proved that houses caused most of the smoke problem. As the obstetrician Lawson Tait exclaimed, "It is merely a popular delusion that the impurity of the atmosphere is attributable chiefly to the tall chimneys of the manufacturers. A summer Sunday's experience at Birmingham will at once remove the fallacious impression. The factory chimneys are not smoking, but the Sunday dinners are being cooked, and the city is as smoky as ever."[19] In 1882 the journal *Nature* held that smoke was primarily caused by domestic fires.[20] The *Encyclopædia Britannica* expressed a similar view five years later, claiming that "domestic fires are mainly responsible for the smoky condition of the atmosphere of our towns; and they for the most part continue to evolve smoke undeterred by legislation or scientific invention."[21]

Lecturing in Manchester in the early 1890s, the antismoke champion Thomas Coglan Horsfall argued that, contrary to popular belief, most smoke came from houses, not factories, and that house smoke was more oily than industrial smoke. He asserted that it was unreasonable to force factory owners to limit their production of smoke as long as houses—the prime culprits—were not subject to any regulation.[22] In 1897 a commentator from Leeds declared that "more odium was cast upon manufacturers than they deserved in the matter of producing black smoke. Statistics showed that on a Sunday, when the factory chimneys were quiescent, the impurities in the air were only 25 per cent less than on week days."[23] Five years later, the *Lancet* expressed a similar view, declaring that "a factory chimney will produce a much more imposing volume of black smoke than a domestic chimney, but the numerical proportion of domestic chimneys to factory chimneys is out of all proportion large, and it is unjust to saddle the factories with all the damage that is done to our health, our public buildings, our works of art, and our gardens by coal-smoke."[24] Continuing along this line in 1909, the *Brewers' Journal* argued self-servingly that "the abatement of smoke from factory chimneys would not put an end to the nuisance. It would undoubtedly diminish it, but its entire abolition in large towns will never be accomplished unless an equally strenuous endeavour be made to diminish smoke from the chimneys of private houses, for the present open fire, pleasant and cheery though it be, is essentially dirty."[25] Similar claims came from Glasgow, where the chairman of the Air Purification Sub-Committee of the city council asserted in 1912 that industry accounted for only a quarter of the smoke in that city.[26]

This rhetoric sought not only to exonerate factory owners for smoke but to blame it on the working class. James Niven (1851–1925), who served as Manchester's medical officer of health from 1894 to 1922, claimed in the early twentieth century that "the biggest evil in connection with smoke is from the cottages."[27] Cottage dwellers were indeed more numerous than those who lived in grander houses, but their per capita consumption of

THE SMOKE PLAGUE: SUNDAY, 11 A.M.

THE SMOKE PLAGUE: MONDAY, 12 NOON.

Figure 7.1. As this 1920 photograph from Sheffield makes clear, the air of cities on Sundays, when factories were idle, was often much clearer than the air on workdays. From William Blake Richmond, "The Smoke Plague of London," in *London of the Future,* ed. Aston Webb (1921), facing 262.

coal was much lower than that of the middle and upper classes, who could afford to consume much more coal. The first Duke of Westminster, Hugh Lupus Grosvenor (1825–99), despite being a leading proponent of smoke abatement, was anything but frugal in his own consumption. Eaton Hall, his country house near Chester, used over two thousand tons of coal a year. To bring in this large quantity, he built across his land a narrow-gauge train line that connected to the national rail network.[28]

The scholars Peter Stallybrass and Allon White have argued that social identity has both a social and a bodily dimension. The Victorian bourgeoisie, in an effort to convince themselves and others that they belonged to "a different, distinctive, and superior class," embraced Descartes' mind-body dualism as a metaphor for social relations. To be respectable, in other words, was to be governed by reason rather than by biology. Members of the middle class regarded their own bodies as an embarrassment, something to be hidden from view and rarely discussed. Instead, they sublimated their anxieties about the body into an "obsessive preoccupation" with the poor, "a preoccupation which is itself conceptualized in terms of discourses of the body." Central to this project was the exclusion of filth. In the minds of many middle-class individuals, assert Stallybrass and White, workers and the poor became synonymous with dirt.[29] To get away from filth, members of the middle class had to do more than simply distance themselves from bodily wastes (a task that the vast sewerage schemes of Victorian Britain had been designed to accomplish). They had to distance themselves from technological filth, including the smoke produced by the steam engines and furnaces of industrial society.

Whether pollution was conceived of as the product of biology or technology, members of the middle class often sought to obscure their own role in its creation. As one factory inspector explained in 1866, the clouds of coal smoke that "may be seen to issue from factory chimneys are for the most part due to the carelessness of stokers."[30] The Birmingham sanitary committee passed a resolution in 1875 that allowed owners of polluting factories to shift the blame to their workers. Smoky chimneys, the committee implied, did not necessarily indicate faulty or inadequate technology; they might be caused by ignorant or inattentive workers. With this in mind, the committee urged factory owners to "deal with the firemen in cases where the fault rests with them."[31] Leeds also adopted policies that held workers rather than factory owners responsible for violation of pollution statutes. In 1898 stokers in that city were served with five times as many notices as their bosses for smoke violations.[32] In Manchester, however, both nuisance authorities and clean air activists long maintained that it was unjust to put all the blame on workers. Speaking to the Manchester and Salford Noxious Vapours Abatement Association in 1881, the factory owner Herbert Philips declared that most industrial smoke was caused by "deficient boiler power and over that firemen had no control."[33]

The assertion that smoke could be prevented if stokers took greater care in loading coal into furnaces had been around for a long time, as had rebuttals to it. An anonymous tract in the 1850s denounced the view that stokers were lazy and asserted that their work—even without the extra effort needed to minimize smoke—was

> as exhausting an employment as a strong and healthy man can well endure; and if to that were superadded the necessity of pushing forward a heavy mass of blazing fuel every ten or fifteen minutes, two men, at least, would be required to do what the superficial examiner of the matter supposes to be attainable by one. Add to this, want of space to work in, and the still greater absence of ventilation, in hundreds if not thousands of the stoking-holes in London, and the astonishment should be how these poor fellows perform their work so well . . . as they do![34]

Despite such arguments, many critics continued to blame workers for the smoke that came from factory chimneys. The trade journal *Chemical News* maintained in 1860 that "it is in vain to expect that the present race of ordinary stokers will devote sufficient care and attention" to prevent smoky fires.[35] In 1902 the Sheffield smoke inspector William Nicholson similarly attributed industrial smoke to "the lazy and slothful fireman," who "will resort to all kinds of dodges and devices to escape a little labour. Self-ease and rest are his constant thoughts; he rests, and is tired, always tired, 'born tired,' and never would be rested even if he had nothing to do."[36] If such fatigue did exist in some furnace rooms, it may have resulted not from indolence, but from carbon monoxide poisoning.[37]

Many commentators argued that because it was possible to burn coal without producing smoke, those who persisted in creating it ought to be held accountable. This line of reasoning might be expected to lead directly to the owners of offending factories, but this was not the case. In 1881, for example, *Chemical News* sought to divert attention from industrial smoke by suggesting that the primary problem came from private houses. Pushing its ideological position even further, the journal blamed smoke on the "trades-unionist carpenters" of London, who had allegedly put up poorly constructed and drafty houses that could be heated only by burning excessive quantities of coal.[38] Others came up with less biased explanations. William Nicholson, the chief smoke inspector for Sheffield, argued that one important factor was the type of coal burned. Bituminous coal, he noted, produced much more smoke than did higher grades of fuel, such as anthracite coal. The choice of which type of coal to use belonged to householders and manufacturers alike. Nicholson also argued that additional smoke resulted from miserly factory owners who overworked their steam boilers in an attempt to obtain more power than they were designed to give. Yet, as noted above, Nicholson blamed incompetent workers for causing smoke. In 1902 he declared that an improvement in the moral character of furnace operators would lead to a reduction of "at least 80 percent of the smoke which is now being belched into the atmosphere." The men currently employed at such jobs, he argued, were ignorant of scientific techniques and exhibited an obstinate "inattention to instructions."[39]

Even when members of the upper classes admitted that their chimneys added smoke to the air, they often denied responsibility for it. Taking their cue from industrialists who blamed apathetic or ignorant employees, landlords and middle-class homeowners often attributed the smoke that came from their property not to poorly designed or inadequate appliances, but to inattention or laziness on the part of their tenants and servants. In 1882, for example, Peter Barlow claimed that the stubborn refusal of "occupiers and domestic servants" to adopt new methods of heating and cooking constituted a major impediment to smoke abatement.[40]

Networks of Reform

During the late nineteenth century reformers in Britain founded a number of pressure groups that sought to limit the impact of industrialization, urbanization, and unfettered commercialism on Britain's natural and built environments. The Commons Preservation Society (established in 1865) challenged landlords' attempts to destroy hedges and restrict public access to rural land;[41] the Society for the Protection of Ancient Buildings (founded in 1877) worked to prevent churches, cathedrals, and other old buildings—tangible symbols of a rapidly receding past—from both overzealous "restoration" and the corrosive effects of air pollution;[42] the Metropolitan Public Gardens Association (established in the early 1880s) worked to increase the amount of accessible green space in London;[43] the

Manchester Field-Naturalists and Archaeologists' Society studied air pollution and promoted both smoke abatement and urban parks;[44] the Scapa Society (begun in 1893) sought to restrict commercial advertising;[45] and the National Trust (established in 1895) dedicated itself to protecting "places of natural beauty or historic interest."[46] The late nineteenth and early twentieth centuries also witnessed the emergence of several groups that attacked air pollution, including the Manchester and Salford Noxious Vapours Abatement Association (established in 1876), the National Smoke Abatement Institution (begun in 1882), the Coal Smoke Abatement Society (formed in 1898), and the Smoke Abatement League of Great Britain (started in 1909).[47]

Lawrence Goldman argues that the members of nineteenth-century voluntary societies created a dense network of "interconnected worlds of middle-class reform."[48] Individuals and groups that worked to reduce the smoke problem were part of this web, and they approached the issue of smoke from a variety of perspectives, such as health, urban planning, social reform, nature preservation, and culture. Important connections existed between smoke abatement, historic preservation, and nature conservation.[49] At the institutional level groups frequently cooperated in support of each other's goals. Advocates who worked to preserve parks and public footpaths, for example, frequently declared that enlarging urban green space would help to purify air polluted by smoke, and historic preservationists alleged that the soot and sulphur dioxide produced by coal combustion were begriming art and architecture and causing stone to erode. Smoke abatement activists supported preservationists by pressuring local officials to banish smoke near Canterbury Cathedral and other historic sites.[50]

Thirty years after the Great Exhibition of 1851, not far from the site where the Crystal Palace had once stood, another exhibition took place in London. In contrast to the earlier exhibition, which had celebrated the promise of industrialization and technology, this one focused on a highly visible downside of the age of steam: the smoke caused by coal combustion. The exhibition was the brainchild of two remarkable individuals, Ernest Hart (1835–98) and Octavia Hill (1838–1912).

A talented intellect, Hart won a scholarship to Queen's College, Cambridge, at the age of thirteen, but decided not to attend because of the discrimination that he would have faced there as a Jew. In lieu of Cambridge he studied medicine in London at St. George's Hospital and at Samuel Lane's Medical School. After working as a surgeon for about a decade,

Hart joined the editorial staff of the *Lancet* in 1863. Three years later, he moved to the *British Medical Journal,* where he served as editor for the rest of his life. While working for the *British Medical Journal,* Hart founded and edited two other journals, the *London Medical Record* (set up in 1873) and the *Sanitary Record* (established in 1874). In addition, he chaired both the parliamentary bills committee of the British Medical Association and the council of the National Health Society.[51] The latter group, founded in 1872, sought to propagate the idea that fresh air, open spaces, and sanitation were essential to health.[52] To disseminate this message, the group organized a team of volunteer lecturers and published a number of inexpensive tracts that emphasized the importance of fresh air to health. As part of its interest in pure air, the National Health Society also lent support to Samuel Barnett's efforts to give urban children an exposure to country life. In 1880, for example, it helped to publicize the search for families who might provide temporary homes, as well as for donors "who would be willing to pay for a few weeks' country air for the little children of our overcrowded London courts."[53]

Unlike many other reform groups at the time, the National Health Society welcomed women into leadership positions. Although its president (the Duke of Westminster) was male, it had several female "patronesses"— including Princess Louise, a daughter of Queen Victoria. Women outnumbered men by more than two to one on its executive committee. In contrast to its welcoming attitude toward women, the society was restrictive in terms of class. Annual membership cost a guinea (21 shillings, equivalent to an entire week's wages for a laborer).[54]

One of the most important environmental activists of the late nineteenth century—and perhaps the most influential woman of that period—was Octavia Hill. Although she is chiefly remembered today for her leadership as a housing reformer and as one of the founders of the National Trust, her influence was much more extensive than that.[55] Helped financially by John Ruskin, Hill set out to prove that landlords could provide decent and affordable housing to working-class families while still earning a profit.[56] In keeping with this capitalistic approach to social reform, many believed that class conflict could be mitigated more effectively by prettifying slums than by redistribution of income. As Samuel Barnett controversially put it, "It is not the poverty that is such a weight upon everybody in the East End, it is the ugliness."[57] Hill and her sister Miranda fully agreed. In 1876 Miranda addressed Hart's National Health Society to announce that she was starting a group dedicated to bringing "art and beauty to the masses." It was called the Kyrle Society, named in honor of the English philanthropist John Kyrle, better known as the "Man of Ross" (1637–1724). The society initially directed its energies toward relatively superficial efforts at making cities more pleasant: it donated flowers to poor people, planted gardens, organized choirs, and put up brightly colored inspirational mottoes on walls. In May 1879 the group organized an open spaces committee, led by Octavia Hill, that fought to turn old urban graveyards into small parks where people could spend time amid grass and flowers.[58] Describing the need for such places in cities, she declared, "There are indeed many good things in life which may be unequally apportioned and no such serious loss arise; but the need of quiet, the need of air, the need of exercise, and, I believe, the sight of sky and of things growing seem human needs, common to all men, and not to be dispensed with without serious loss."[59] Hill's work on this committee united her interests in housing reform, nature, and improving the air that people breathed.[60] In her words, smoke abatement was "a natural sequel to what we had tried to do to secure open spaces," so that cities could be places where "trees and flowers and grass would grow in and show their natural colour and brightness."[61]

Ernest Hart and Octavia Hill knew each other long before they became leaders of the smoke abatement movement. They probably met in 1873 at the wedding of Henrietta Rowland and Samuel Barnett, both of whom had aided Octavia Hill's philanthropic efforts to improve workers' housing in London.[62] One year later, Hart successfully interceded on Hill's behalf when one of the properties she managed in Marylebone was threatened with demolition by vestry officials.[63] If the two had not yet become well acquainted, they certainly were so by the end of this episode. As Miranda Hill wrote in November 1874, "Mr. E. Hart . . . looked into the matter thoroughly with O. and said he thought she had a *very* good case."[64]

Persuasion

In the early months of 1880, following a winter of deadly smoke and fog in London, Hill and Hart crossed paths in the Mediterranean. Hart and his wife Alice, who were traveling with Samuel and Henrietta Barnett, sailed to Greece in early April, where they met up with Octavia Hill and Harriot Yorke (1842 or 1843–1930), who remained her partner until Hill's death in 1912.[65] Although it is not clear how much time Hart and Hill had together,

Figure 7.2. Octavia Hill, ca. 1880. From C. Edmund Maurice, ed., *Life of Octavia Hill, as Told in Her Letters* (1913), frontispiece.

Figure 7.3. Ernest Hart, ca. 1870s. Photograph by George Robert Fitt. Reproduced courtesy of the Wellcome Trust Medical Photographic Library (V0026517).

it seems likely—based on the conditions of the previous winter and their subsequent attention to the smoke issue—that they discussed London's coal smoke. Smoke was certainly on Hill's mind in the following month when she visited Nuremberg. In a letter to her mother Hill observed, "Trees grow among the houses; and children play round them, and clean, industrious women knit at their doors . . . ; these gardens for the people everywhere look reproach on me when I think of England, and every tree and creeper and space of green grass in the town reminds me of our unconsumed smoke and how it poisons our plants, and dims the colour of all things for us."[66]

When Octavia Hill returned to England in the summer of 1880, writes one of her biographers, "her first enterprise was the struggle against smoke, to which she had vowed herself in Nüremberg [*sic*]. . . . Accordingly, as soon as she got home, a subcommittee of the Kyrle Society was formed to deal with the subject of smoke abatement."[67] In July Hart and Hill asked their respective societies to form a joint committee to address the problem of fog and smoke. The National Health Society provided office space to the committee and gave it a modest grant of ten guineas. One of Hart's first actions as chair of the new committee was to ask Sir William Chandler Roberts (1843–1902), professor of metallurgy at the School of Mines and chemist to the mint, to assist the new group, a request that he accepted.[68] In October 1880 the Fog and Smoke Committee, as the group was first known, decided to organize an exhibition of smoke-preventing appliances.[69] The committee had powerful connections, and the lord mayor of London soon offered to make the Mansion House, his official residence, available for the group to hold a public meeting.[70] These developments received extensive press coverage. Applauding the work of the committee, the *Lancet* declared that polluted air had become a "grave national calamity and one which calls for the active interference of the legislature."[71] Even the industry periodical *Chemical News* expressed support, noting that "a new phase of what is called 'sanitary reform' has just been brought forward. We have been told that the fogs of London and of our other large cities are a nuisance, and though the news is somewhat of the stalest, we are listening as if to some novel revelation and beginning to look out for a remedy."[72]

Hill and Hart believed that technology had the potential to cause serious social and environmental harm, but they were also convinced that it held the solution to such problems. They had two goals in organizing the exhibition: to increase public attention to air pollution and to demonstrate that smoke could be prevented. Embodying a quintessentially Victorian faith in progress, they argued that cleaner air required neither government compulsion nor the abandonment of industrialism. Coal combustion need not pollute the air; all that was needed was information and education.[73] Douglas Galton, who later served as a vice-president of the National Smoke Abatement Institution, also insisted that reform should occur through voluntary efforts rather than government pressure. "The purification of the air of London from smoke," he asserted in 1880, " . . . cannot be obtained by legislation alone; legislation on such matters must be accomplished by the co-operation of the individual members of the community. When that co-operation has been secured by the spread of sanitary knowledge, we may hope for practical progress."[74]

While leaving the door open for government regulation of industrial smoke, Galton unequivocally rejected legislative action to reduce smoke from private houses. The question of convincing people to stop using smoky open fireplaces, he insisted, "seems scarcely one which can be dealt with by legislation."[75] Education was his preferred approach, and most other reformers in the 1880s agreed. Hart shared the view that education and persuasion were preferable—at least for the time being—to legislative fiat. He "did not believe they could cure everything by putting into acts of Parliament 'must' or 'shall' or 'may.' He did not think public opinion would bear that out, and his own experience was that the introduction of imperative words such as 'shall' was apt to lead to a reaction, thus destroying the very objects they desired to meet."[76] The notion that individual householders were responsible for much of London's smoke problem was embodied in the 1881 Smoke Abatement Exhibition, which aimed at reforming not only industrial coal use but also that within private houses.[77] Hart declared that "at a moderate cost and with ordinary intelligence and goodwill every one may fulfill his part in lessening the pall of smoke and of unconsumed carbon which now daily overhangs London."[78]

The Fog and Smoke Committee adopted a conciliatory attitude toward industry. As one of its members put it, they should do everything possible to avoid interfering with the "manufacturing interests of the metropolis, the importance of which the societies recognized."[79] In November 1880 the group held a conference intended to enlist manufacturers in the movement for smoke abatement. Hart opened by expressing his wish that progress could be made "with the least possible friction and with the largest possible assistance" from manufacturers. The committee hoped that factory owners would prevent smoke if they were shown how

Figure 7.4. "Medal to commemorate the Smoke Abatement Exhibition. Dedicated to the Grate Coals (without smoke), of Kensington." Shortly after the smoke abatement exhibition opened, Harry Furniss contrasted the smoky atmosphere of 1881 with conditions that might prevail a year hence, when thanks to the efforts of the reformers, the sun would shine, flowers would bloom, and Cleopatra's Needle and St. Paul's Cathedral would be free from smoke. From *Punch*, 17 Dec. 1881, 281.

to do so, since efficient combustion would use less coal and thus save them money.[80] Organizers of the 1881 Smoke Abatement Exhibition explicitly rejected the view that industry accounted for the vast bulk of the smoke problem. Hart, for one, claimed that most of the smoke in London came from "kitcheners" (cooking ranges).[81] To solve the smoke problem, they believed, millions of people had to change the way in which they used coal.

The exhibition opened on 30 November 1881 with a ceremony at the Royal Albert Hall hosted by the lord mayor of London and attended by a number of prominent aristocrats, including Princess Louise (1848–1939), who would later play a leading role in the National Trust and the Coal Smoke Abatement Society, members of both houses of Parliament, and prominent scientists and engineers.[82] The exhibition occupied rooms attached to the conservatory at the rear of the Royal Albert Hall, where visitors were confronted by exhibits of smoke-abating appliances from over 230 British and American firms.[83] Reflecting the view that the smoke problem was caused by both domestic and industrial combustion, the exhibition featured equipment ranging from household stoves and grates to industrial boilers, which were tested to determine how effective they were in preventing smoke. Once people had this information, the organizers assumed that they would act rationally and adopt the technologies that were least polluting. The organizers maintained that it was not necessary to rely on people acting with the good of the community in mind; self-interest alone would lead people to choose the cleanest technology because smoke was wasted fuel.[84]

The exhibition proved an immediate success. Thirteen thousand people visited it during the first week, and attendance continued to exceed expectations thereafter. To accommodate the large number of visitors and allow people with busy schedules to attend, the hours of opening were extended until 10 P.M.[85] The exhibition gave many people their first glimpse of gas heating and cooking, but the impression that they took away may not have been as positive as Hill and Hart would have wished. Fireplaces and stoves designed to reduce the amount of smoke entering the atmosphere sometimes created a problem of "noxious fumes" in the rooms where they were used. *Nature* reported that the section of the hall where gas appliances were housed was filled with "the very unpleasant fumes of burnt gas—caused, we suppose, by so many of the gas fires shown not being provided with flues." This odor, it noted, "must prejudice visitors very much against the use of these very valuable appliances."[86]

The exhibition received extensive coverage in the newspaper and periodical press. In an apparent effort to dispel the view that smoke abatement was a frivolous fad, the *Builder* praised its "earnest and business-like character." Describing the scene inside, it reported that visitors took the subject equally seriously: "Here you see the country clergyman earnestly studying a stove which in his mind's eye he has already fitted into his carefully-restored church. There is the lecturer, the sanitary reformer, the medical man, who is chiefly bent on studying the union of the difficult aims of warming and ventilation. Through all these thronging groups of reasonably anxious faces peer the keen eyes of the salesmen, the inventors, the speculators."[87] Although the *Builder* welcomed the exhibition, the journal argued that education and persuasion alone were insufficient. The only thing that would significantly reduce the amount of smoke in the air of London, it predicted, would be a complete ban on fuels that produced smoke.[88] In contrast to this favorable response, the *Lancet* reacted coldly when the exhibition began, asserting that although it would "doubtless attract some attention and do a certain amount of good . . . , its practical bearing on the question of fogs and the smoke nuisance is somewhat vague."[89] Two weeks later, however, the journal's tone toward the exhibition was markedly warmer. Perhaps having been chastised by one of the

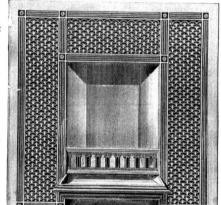
Figure 7.5. Manufacturers of "smokeless" appliances quickly capitalized on the smoke abatement exhibition in an attempt to boost sales. This advertisement appeared in a book published by the very group that had awarded the prizes. From *Report of the Smoke Abatement Committee, 1882* (1883).

influential supporters of the Smoke Abatement Committee (as the group was now known), the *Lancet* did an about-face and declared that the exhibition "has most deservedly attracted a large share of public attention, and we most heartily wish success to this effort to diminish what has really become a national nuisance. It is hardly necessary to dwell upon the practical evils which result to this overgrown metropolis from its smoke-laden atmosphere."[90]

The smoke abatement exhibition at South Kensington ran for two and a half months. By the time that it closed in February 1882, about a hundred thousand people had seen it, many of them from places far from London. The committee had worked hard to publicize its activities and had asked philanthropic groups and local authorities in large towns and cities throughout Britain to provide financial support to the exhibition.[91] Although it is not clear how much money this appeal generated, the committee succeeded in gaining widespread attention. The municipal authorities of several leading industrial cities, including Sheffield, Manchester, and Birmingham, sent representatives to view the exhibition.[92] Official delegations also came from many of the world's major coal-consuming countries, including France, Germany, and the United States.[93]

Shortly after the exhibition closed, its organizers and influential supporters reconvened at a dinner party hosted by the lord mayor of London at the Mansion House. The guest list included the writer George Augustus Henry Sala; the U.S. ambassador to Britain, James Russell Lowell; several members of Parliament; the mayors of Manchester and the nearby towns of Stockport, Bolton, and Oldham; the directors of the South Kensington Museum and the National Gallery; and the presidents of the Royal Academy of Arts, the Institution of Civil Engineers, the British Association for the Advancement of Science, and the Royal Society.[94]

Members of the Manchester and Salford Noxious Vapours Abatement Association were so impressed that they asked the Smoke Abatement Committee for permission to take the exhibition to Lancashire in the spring of 1882.[95] Like its predecessor in London, the Manchester exhibition showcased a large number of "smokeless" stoves, boilers, and other devices for domestic and industrial use.[96]

After the exhibitions in London and Manchester ended, members of the Smoke Abatement Committee voted unanimously to reconstitute the group on a more permanent basis as the Smoke Abatement Institute,[97] which soon evolved into the National Smoke Abatement Institution. To emphasize the importance of the work in which the group was engaged, its 1884 report boasted that "the United States government, who [*sic*] sent a special commissioner to the Smoke Abatement Exhibition and printed his report as a state document, have recently applied for additional copies of the reports and publications of the Institution."[98] Despite the encouragement that this international attention provided, by the middle of the 1880s the National Smoke Abatement Institution had lost much of its early enthusiasm. Many members of the group were discouraged at the seemingly intractable nature of the smoke problem, and some, including Hill, turned their attention to other issues.

While many public health experts supported the goals of the new smoke abatement group, some feared that the attention it was getting might leave their own activities in the shadows. In 1883 Alfred Carpenter, a vice-chair of the Sanitary Institute, complained to Edwin Chadwick that "the antagonistic influences" of the "National Health Society, Smoke Abatement S[ociety], and Sanitary Aid S[ociety] are all diminishing our power as sanitarians. I want to get an amalgamation of all these into one body and want to get ourselves before the [Local Government Board]." Not surprisingly, Carpenter suggested that the Sanitary Institute would be the ideal umbrella for such a plan.[99] Chadwick, it should be noted, was a vice-president of both the Sanitary Institute and the National Health Society during this period. Whether as a result of Carpenter's machinations or not, the latter organization severed its relationship with the National Smoke Abatement Institution in 1884. The annual report of the National Health Society for that year made it clear that its leadership no longer wished to devote any time to the issue of coal smoke: "Members or others who have any communication to make on the subject of 'Smoke Abatement,' should communicate with Mr. E. White Wallis, Secretary to the [National Smoke Abatement] Institution." The address given was that of the Parkes Museum of Hygiene, which was connected to the Sanitary Institute. Subsequent reports of the National Health Society said nothing about the National Smoke Abatement Institution.[100]

Compulsion

The National Smoke Abatement Institution became less active in the late 1880s, in part because it leaders disagreed about how to proceed. In contrast to his earlier position, Hart came to believe that society had a strong interest in compelling individuals to prevent smoky fires on public health grounds. Testifying before a parliamentary committee in 1887, Hart declared that "the increase in the volume of smoke emitted in London within the last twenty years has produced perceptible differences in the health of the people. . . . [A]ll of the vital processes of healthy life are allowed to droop for lack of the actinic rays of the sun; they are struck down by the increasing volume of smoke, which means a general deterioration of the health of the children and grown-up people also." The advantages to be gained through smoke prevention, he argued, were "great to the community at large, but the individual sees his present inconvenience and does not see his individual profit."[101]

Those who continued to be active in the smoke abatement effort found themselves divided about how to proceed. Some remained opposed to increasing the government's role, while others believed that the group had devoted too much attention to education and not enough to legislative reform. In a move designed to lay a foundation for such reform, the group called on Gladstone's administration to appoint a royal commission to investigate the technical and legal aspects of smoke from both domestic and industrial coal use.[102] These efforts collapsed when the home secretary rejected the proposal. Although the government eventually appointed a select committee to look into the matter, this was merely a token gesture. The committee called few witnesses, and its report soundly failed to provide the impetus for further legislation.[103] In 1884, disenchanted with the group's failure to achieve meaningful progress through education alone, the Scottish Conservative Lord Stratheden and Campbell (William Frederick Campbell, 1824–93) introduced a bill that would have regulated household smoke. It was defeated. He persisted in trying to get the measure passed during the following nine years, but each time he introduced it, other MPs decried it as an unjustifiable attack on the idea that the home was one's castle.[104]

Although the members of Parliament often managed to quash proposals to enact further restrictions on smoke, they were less successful in preventing London's air pollution from entering the legislative chambers. According to one wag, the poor quality of the air in the House of Commons might explain the poor quality of the rhetoric there, compared to the engaging speeches that MPs generally delivered "at social gatherings and public dinners. . . . Can it be," the writer asked facetiously, "that the difference is only a matter of ventilation?" Considerable sums were spent on efforts to supply the chamber with pure air, but success proved elusive. Cotton wool filters, installed in the inlets to the ventilation system, quickly became clogged with soot and smoke. After only a short time in place, they were "rendered black, utterly filthy, and soiling to the touch." Reporting in 1889 on yet another unsuccessful attempt to purify the air entering the House of Commons, the *Lancet* asserted that smoke abatement activists were correct in claiming that the problem required a broader solution. Few individuals, it declared, could follow the approach taken by Parliament and install filters on their windows; "even if this were possible, our streets would be no better." Instead, politicians ought to realize that "if they determine to legislate in a clear atmosphere, they must provide for the prevention of fog in the whole of London as well as in the legislative chamber itself." Ten years later, during another episode of unusually thick air pollution in London, one MP complained that "the committee rooms, the terrace, and the dining rooms were filled for hours with most unpleasant fumes."[105]

Despite the waning of antismoke activity in London, new efforts to reduce smoke began in industrial cities in the late 1880s. Reacting against what he characterized as "the *dilettante* smoke abatement agitators" in London, who had "made themselves ridiculous and damaged their own cause by the extravagance of their assertions respecting the money benefit of smoke-preventing appliances," the Lancashire councilman (and coal mine owner) Herbert Fletcher argued that efforts to reduce smoke should be led by scientific experts.[106] This concern stimulated the formation of the Committee for Testing Smoke-Preventing Appliances. The group was led by Alfred Evans Fletcher (1826–1922), a chemist who had succeeded Robert Angus Smith as the head of the Alkali Inspectorate in 1884. At its first meeting, held in Manchester's town hall on 8 November 1889, attendees passed a resolution declaring that factory smoke resulted from the absence of "accurate information" rather than from indifference or greed.[107]

In contrast to the disdain with which some critics had viewed earlier attempts to evaluate smoke-preventing technology, this group was praised—even by the industry journal *Engineering*—for work that was "authoritative and unquestionably impartial."[108] The group received much of its funding from manufacturers, who had to pay a membership fee of one-half pound for each boiler they used.[109] Although the committee worked to distance itself from earlier efforts, there was a great deal of continuity. In addition to factory owners, its membership included many who had long been active in smoke abatement, such as the Duke of Westminster, Alfred Carpenter, Douglas Galton, and Fred Scott.[110] Its fundraising techniques were also similar to those used a decade earlier by the organizers of the South Kensington exhibition. In September 1890 the Birmingham health committee donated £10 to the group after learning that the corporations of Manchester, Salford, and St. Helens had each given this amount.[111]

In the steelmaking city of Sheffield, antismoke activism was spearheaded by what Edward Carpenter termed "our little group of Socialists." Not surprisingly, they faced an uphill struggle. Late in his life Carpenter recalled, "We got a table out into the street, made little speeches, and distributed leaflets and pamphlets. The workpeople mostly jeered," believing that smoke was an inevitable accompaniment to the industries in which they worked.[112] Despite this opposition, approximately fifty-eight hundred people signed a petition in 1889 calling for stronger enforcement of smoke laws, and the town council responded by creating a smoke nuisances subcommittee.[113] The organizer of the petition, Benjamin Burdekin (1831–1906), helped to form the Sheffield Smoke Abatement Association, of which he became the honorary secretary.[114] As a result of pressure from this group, the permissible emission of black smoke was lowered in 1892 from ten minutes an hour to six minutes, and later reduced further. Although fines remained low, the medical officer of health claimed in 1897 that the average amount of dense smoke had declined in less than a decade from ten minutes an hour to under three minutes.[115] Officials in other industrial cities made similar claims, but not everyone agreed with them. Testifying before a government committee in 1903, Manchester's chief medical officer boasted that the air of his city was "very much less polluted" than it had been formerly, yet Rev. W. E. Edwards Rees argued that it was "worse than ever. You have stopped not all, but some of the most flagrant abuses, but the amount of abuse is greater than ever it was."[116]

Building on the work of the Committee for Testing Smoke-Preventing Appliances, a new group began in the mid-1890s with the intention of coordinating the efforts of smoke abatement societies in Sheffield, Middleton, and Manchester. The league enjoyed particularly strong connections to Manchester reformers, and its secretary was the ubiquitous Fred Scott.[117] The group held a huge gala outside Manchester in 1895, attended by five thousand people. Susan Hopwood, who hosted the event, was president and secretary of a branch of the Smoke Abatement League in Middleton, a place where local authorities showed little willingness to prosecute the owners of smoky factories.[118] Referring to her work, one of her contemporaries asserted that "in the thickly-populated district between Oldham and Manchester it is left to a lady to fight almost single-handedly the battle against the local mill-owners." Hopwood won numerous lawsuits against polluters, but the fines imposed appear to have had little effect in reducing smoke.[119]

Praising the growing agitation against smoke across Britain, the *Builder* commented in 1899 that "in Manchester, Leeds, and other large towns as well as in London there are public-spirited men gathering information on the subject and drawing the attention of householders to the inconvenience, discomfort, and damage inflicted by smoke, and even prosecuting offenders." The journal argued, however, that this was not enough. Further legislation was needed, both to ban smoke from private houses and to transfer the enforcement of industrial smoke from local authorities to officials who could not be induced to look the other way.[120]

Horticulturist F. W. Oliver suggested in 1891 that it ought to be possible to protect plants from the damaging effects of polluted air by enclosing them in a greenhouse that was "absolutely fog-proof, with close glazing, triple doors, and padded ventilators." He added that "filtered air could be supplied, as it is in the House of Commons, by pumping through several inches of cotton wool."[121] Several other buildings throughout the country were fitted with filtered ventilation systems during the late nineteenth century, notably hospitals. In Glasgow the Victoria Infirmary used a large wet screen to filter incoming air. Arguing that this method was much more effective than the dry-cotton filters that had recently been tested in the House of Commons, Galton observed, "We seem to be unable to make the necessary sacrifice of our open fires in order to remove black fogs from our streets; but by the method above described we might at any rate secure a clear and clean atmosphere when indoors."[122]

After a period of little activity in London, smoke abatement forces regrouped in December 1898 to form the Coal Smoke Abatement Society. Like the National Smoke Abatement Institution of the 1880s, its membership was weighted heavily toward the upper reaches of the social scale, with a preponderance of aristocrats on its governing committee. One of the first projects of this new antismoke group was to put pressure on the London County Council to take action against polluters when borough authorities failed to act. In a letter to the *Times* in 1904, a supporter of the group argued that its work was too important to be left solely to private initiative. Borough authorities or the London County Council should assume greater responsibility for controlling smoke, and the latter ought to "appoint a special committee to look into this really important matter."[123]

When nothing came of this proposal, the Coal Smoke Abatement Society continued to use inspectors of its own to bring violations to the attention of local authorities.[124] As its president later recalled, "We reported nothing under ten minutes of Black Smoke in the hour, so timid were we of being dubbed mad faddists."[125] One place in which the new group's efforts appear to have prompted greater enforcement was Lambeth, where the magistrate fined the South London Electric Company £100 in 1901.[126] A short time later the *Lancet* credited the society with inducing many of the local authorities governing London's boroughs to crack down upon smoke.[127] The society claimed that the number of smoke nuisances within the county of London fell from 187 cases in 1900 to only 15 in 1908. Its secretary complained, however, that much of the smoke obscuring London's skies came from factories located elsewhere.[128] The society repeatedly asked the Local Government Board to take action, without success. Frustrated with this inattention, activists hoped to extend the authority of the council so that it could deal with smoke from neighboring areas, but its efforts again ended in disappointment.[129] Reformers from Manchester, Leeds, and Sheffield had no better luck against the "apathy and indifference" of the Local Government Board when they petitioned it in the early years of the twentieth century in an attempt to ensure that smoke regulations were enforced uniformly throughout England and Wales—perhaps by the Alkali Inspectorate or an analogous national entity.[130]

Although the government failed to satisfy these activists, it showed signs of increased interest in the problem. In 1904 the Foreign Office asked British ambassadors stationed in Washington, D.C., and several European capitals to furnish information about how other industrialized countries regulated smoke. The ensuing report, presented to Parliament in 1905, focused particular attention on the United States; it even included a copy of the legal notice that Chicago's chief smoke inspector issued to violators of the city's smoke ordinance.[131]

Many of the central figures in the Coal Smoke Abatement Society—including Sir William Blake Richmond, Rollo Russell, and the Duke of Westminster—had taken part in the National Smoke Abatement Institution of the 1880s and supported Lord Stratheden and Campbell's attempts to enact a more stringent antismoke law.[132] In 1883 Westminster pledged £500 to the National Smoke Abatement Institution.[133] Westminster supported Richmond's antismoke efforts in the late 1890s by hosting public meetings of the Coal Smoke Abatement Society and by giving money to the cause of clean air.[134] Long after the death of the first Duke, his family continued the support that he had begun by providing free office space to the Coal Smoke Abatement Society.

Figure 7.6. Wielding a sword and a palette, the artist Sir William Blake Richmond, founder of the Coal Smoke Abatement Society, battles industrial smoke. Cartoon by E. T. Reed in *Punch*, 3 Dec. 1898, 262.

In contrast to the National Smoke Abatement Institution, which had devoted considerable attention to household smoke and relied on gentle persuasion as its primary tactic, the Coal Smoke Abatement Society believed that the focus should be on industrial smoke and that more aggressive enforcement was necessary.[135] The question of whether industry or private houses produced more smoke had important political ramifications. Factory owners naturally sought to divert attention from their own smokestacks, and the argument that industry contributed relatively little to the overall smoke problem helped them to do so. Antismoke activists also engaged in political calculation. Although reformers like Richmond recognized that domestic fires were a major source of smoke, many of them believed that they would achieve more by concentrating on industrial smoke. Despite attempts to pass laws restricting household smoke, the enactment of such legislation on a national level would not occur until after the London smog disaster of 1952.

The Coal Smoke Abatement Society received frequent assistance from the *Lancet,* which provided ample and favorable coverage in its pages and helped the group to evaluate the claims about smoke-preventing fireplaces. This built on earlier work done by the *Lancet* in the 1890s, in which its own Analytical Sanitary Commission had conducted a detailed series of tests involving smoke-preventing appliances.[136] The joint editors of the journal, Thomas Wakley Jr. and Thomas Wakley Sr., both sat on the governing committee of the Coal Smoke Abatement Society.

Richmond's approach represented a fundamental shift in environmental activism, one that was in tune with a changing political climate. In contrast to the laissez-faire principles that predominated during the mid-nineteenth century, by 1900 a growing number of people had begun to reject the notion that the quest for private gain inevitably promoted the good of society. Instead, they came to believe that the power of the state could and should be used to expand individual liberty in many areas previously left to private initiative. Things that had not previously been a "right" became so, in such areas as education, housing, and insurance. In 1888 Herbert Fletcher argued that pure air was a similar right. The British people, he asserted, who had "striven for a juster distribution of power and the establishment of many a legal right during the century," ought to claim "their natural right to see the sun, when he shines, and to breathe the air clean," at least to the extent that this was "compatible with the industrial prosperity of the country."[137] Echoing Fletcher's suggestion that state intervention ought to be understood as a tool for expanding rights rather than as something that constrained them, one smoke reformer asserted in 1901 that "an individual who is one of a community . . . must not do that which will produce ill effects on the health and comfort of the community."[138]

Supporters of industry responded with hostility as soon as they heard that a new antismoke group was forming. In December 1898 the *Engineer* rejected the idea that London could be made smokeless as a "pretty excursion into dreamland," and asserted that factories and clean air were incompatible. Arguing that a city without smoke would also be one without jobs, the journal declared, "Ideal London might be a very terrible place in which to live." It also suggested that smoke prevention was a socialist idea, for a government interventionist enough to prohibit smoke would likely pursue other unwelcome policies, such as the provision of public housing "in great blocks of buildings, with common libraries, baths, and playgrounds." To get rid of smoke Londoners would have to sacrifice their factories, their fireplaces, and their individual freedom.[139]

One month later the trade publication *Industries and Iron* launched an even more direct attack against the new group, even though it could not be bothered to identify it correctly: "We find that the National Association for the Abatement, or Suppression—we do not quite recollect which—of Smoke is still engaged in its philanthropic labours. The president of the association is the Earl of Meath [Meath was in fact one of several vice-presidents; Richmond was its president] who has already won a liberal meed of recognition as the champion of 'open spaces' in London districts." The article went on to assert that the group's decision to focus on London was misguided, as the smoke problem was more serious in industrial regions. "To confine the benefits of their work to the personal area of most of the members, appears a little selfish on the part of the association. We make this suggestion in good faith, because admittedly the smoke nuisance is much worse up North than in Southern districts." In a final attempt to belittle the group and undermine its claims to speak for the public interest, the article claimed that the chairman of the group owned a patent on "a smoke-consuming device for attachment to boiler furnaces. A little business mingled with philanthropy is an excellent thing."[140]

Taking up this challenge, the Earl of Meath took the message of the Coal Smoke Abatement Society to Manchester. Speaking there in 1902, he noted,

> Your corporation [city government] has provided you with excellent drinking water. Would that they could provide you with equally pure and wholesome air in pipes from the mountain summits of the Lake district. But if they cannot do this, much may still be accomplished towards purifying the atmosphere, provided only the existing laws on the subject be strictly enforced. London air is already purer, thanks to the exertions of the Smoke Abatement Society . . . which makes its business to see that

the Authorities put in force the laws for the purification of the atmosphere. Such an organisation might not be out of place in your own city.[141]

Meath's visit to Manchester attracted considerable attention and helped to reestablish links, formed in the early 1880s, between antismoke activists in that city and the capital.

Five years after Meath's speech, the Quaker educator John W. Graham published a tirade against smoke entitled *The Destruction of Daylight*. Echoing Marx, he declared that "the pile of the feudal castle has given way to the pile of the modern factory." Arguing that the state had a responsibility to protect the community against "the new feudalism," Graham insisted that "no class has ever been fit to be entrusted with unfettered power; and the modern manufacturing firm or capitalist, wise and good as he often is individually, is no exception to that rule." Everyone had a right to pure air, and the state had a compelling interest in protecting it, even if this meant fines for individual householders.[142] Two decades earlier, such arguments had been rejected as extreme even by people involved in the smoke abatement movement. By the eve of World War I a growing number agreed with Graham, but Parliament refused to pass laws regulating domestic smoke.

In 1909 smoke abatement activists and concerned municipal officials from Glasgow, Sheffield, Manchester, and other industrial cities formed a new group to coordinate the fight against smoke. Their new organization, the Smoke Abatement League of Great Britain, soon sent a delegation to meet with the recently installed president of the Local Government Board, the Labour MP John Burns. William B. Smith, the acting president of the Glasgow and West of Scotland branch of the Smoke Abatement League of Great Britain, told him what had become a common refrain: each city "could only take action with regard to works within its own area, whilst it often happened that there were works just beyond the city or borough which poured out smoke continually, yet no action could be taken."[143]

Reformers from several countries convened in London in 1912 for a conference on smoke abatement. Some speakers at this conference, such as William Nicholson, the chief smoke inspector for the northern city of Sheffield, maintained that further legislation was unnecessary; modern equipment, "scientific feeding" of industrial furnaces, and specialized training for smoke inspectors would be sufficient to solve the smoke problem.[144] Other speakers, however, insisted that it was pointless to rely exclusively on persuasion and enlightened self-interest. Notable improvements had occurred in some places, but numerous towns and cities remained full of smoke. Too many factory owners seemed content to burn coal wastefully rather than invest in less polluting equipment, safe in the knowledge that any fines they incurred would be modest. Local authorities, these critics argued, often failed to enforce smoke laws out of fear that offending industries would move away, taking their tax revenue and jobs with them. To establish uniform and comprehensive standards across the country, immune to local pressure and modification, a growing number began to share Richmond's view that smoke inspection should become a responsibility of national government.[145] Richmond's wish was partially realized in 1958, when the Alkali Inspectorate assumed regulatory control over electric power stations, gasworks, and coke plants, but by that time Britain's industrial supremacy had become a fading memory, and coal smoke had lost its remaining associations with manufacturing prosperity.[146]

CHAPTER 8

Regulating Pollution

The present condition of the air we breathe is nothing short of a grave national calamity, and one which calls for the active interference of the Legislature.
—The Lancet, *1880*[1]

FOR MOST OF THE nineteenth century the right to vote in Britain was restricted to adult male property owners. Women did not gain the right to vote until 1918; even then, it was initially restricted to women aged thirty and over. Both houses of Parliament were dominated by the upper classes, and politicians across the ideological spectrum believed that one of the principal roles of government was to protect private property. As a result, most were extremely hesitant to grant government the power to interfere with the exercise of property rights. Despite this laissez-faire ideology, the authority of government to regulate the actions of property owners expanded significantly from the 1830s onward. What accounts for this apparent paradox? The answer lies in the theory of individualism that formed an integral part of nineteenth-century political economy.

According to this view, those able to make decisions for themselves (mentally competent adult males) should be allowed to use their own property and labor power without the interference of government, as long as they did not impose costs on other individuals who did not or could not consent to them. This logic explains how legislation restricting the actions of property owners could win passage during the mid-nineteenth century. The Factory Acts of 1833 and 1847, which limited the hours of employment, applied only to women and children—individuals who were thought incapable of negotiating terms of employment on their own behalf. In contrast to this, bills aiming to protect adult males from overwork and unsafe occupational conditions were frequently opposed on the grounds that they interfered with the rights of competent individuals to establish their own contractual terms.

Early environmental legislation was similarly justified by the argument that state intervention was necessary in cases where it was impractical for individuals to identify and hold responsible the particular firms that were causing harm to their property. The Alkali Act of 1863, for example, functioned primarily to protect landowners from having their trees, hedges, and agricultural crops damaged by hydrochloric acid emissions from the chemical industry. Supporters of the bill argued that although they would have preferred to see the two sides work privately to negotiate compensation for this damage, such a solution was impossible given the existence of multiple chemical works in many of the areas affected by this problem.

Related to the question of consent was the issue of whether individuals possessed the ability to take reasonable steps to avoid a particular risk or inconvenience. Clean air activists who sought to expand government regulations frequently insisted that people could do little to protect themselves against smoke. Fred Scott, who spent his career running a multitude of public health and antipollution groups in Manchester, asserted in the 1890s that the intervention of government to protect the health of the individual was even more justified in the case of smoke prevention than in regard to water and food. For a man "can boil water or milk before use, and he can have any article of food analysed, . . . but he cannot protect himself against the pollution of air upon which he has more immediately to depend for existence than upon any other commodity." Despite this, argued Scott, local sanitary authorities frequently neglected their duty to prosecute those who polluted the air, and magistrates often failed to impose penalties.[2]

Others shared Scott's view that strong state intervention was needed to prevent air pollution. Admonishing Sir Henry Hussey Vivian, a Liberal MP who believed that the state had no business entering factories to ensure that manufacturers complied with laws regulating air pollution, the

Builder asserted in 1880 that the rights of private property had to be weighed against the rights of all people to common resources such as air. Turning the property rights argument on its head, the journal asserted that the public had "no desire to go into this gentleman's works so long as he does not come into their atmosphere." Surprisingly, Vivian later joined the Smoke Abatement Committee, on which he no doubt exerted his opinion that the way to reduce pollution was through persuasion rather than compulsion. Yet as we will see later, many smoke abatement reformers in the 1880s supported such an approach.[3]

Legislation and Enforcement

For most of the nineteenth century Britain had no national laws limiting the amount of coal smoke that could be emitted by industry or houses. Common law allowed aggrieved parties to sue those who created a nuisance, but the availability of this recourse did little to control smoke.[4] In the 1840s a number of towns and cities included smoke abatement clauses in the improvement bills that they submitted to Parliament. Not all towns adopted smoke legislation, however, and even those that did often treated polluters leniently. The situation in London was particularly complex. London was not a single political entity but rather an agglomeration of special-purpose governing bodies such as vestries, turnpike authorities, sewerage districts, and the metropolitan police district. The "City of London" constituted little more than one square mile within the much larger area comprising the metropolis as a whole, and it jealously guarded its historic independence.

In 1851 the City of London acquired the authority to levy fines on factory owners who filled the air with smoke. This act succeeded in reducing the amount of smoke emitted in the ancient square mile, but it did not apply to the rest of London. This changed two years later when Parliament enacted the Smoke Nuisance Abatement (Metropolis) Act of 1853, which required manufacturers to "consume or burn as far as possible all the smoke" from their furnaces. Enforcement of this law lay in the hands of the metropolitan police. As a result of direct involvement by the home secretary, Viscount Palmerston, the police enforced the law aggressively during its initial years of operation. Between August 1854 and March 1855, 147 cases went to trial, of which 124 resulted in conviction.[5] The 1853 act was strengthened three years later by amendments extending it to steamboats on the Thames and certain industries that had initially been left out.

These acts appear to have led to some reduction in the amount of industrial smoke in the metropolis, but enforcement was uneven, and many smoky factories remained exempt from regulation.[6] By the 1870s London's smoke regulations were generally ignored by polluters and municipal authorities alike.[7] Even when cases went to court and convictions were obtained, fines were ridiculously low—often no more than half a pound and sometimes as little as a penny.[8] Another reason air quality remained poor was that reductions in industrial smoke were more than offset by increases in the average amount of coal each household consumed.[9]

The Alkali Act of 1863 created the first national environmental regulatory body in the world. Its initial mandate applied to only one industrial process, the manufacture of alkali (sodium carbonate), and to only one pollutant, hydrochloric acid vapors. The act was later amended to include other processes and pollutants, including sulphur compounds generated in the production of heavy chemicals. Paradoxically, however, the inspectorate lacked the authority to regulate sulphur entering the air as a result of coal combustion.[10] This led to a bizarre situation in which coal-fired furnaces sometimes emitted more sulphuric acid than was allowed to be released from works manufacturing the chemical.[11] Alfred Evans Fletcher (1826–1922), who succeeded Robert Angus Smith as chief alkali inspector in 1884, sought unsuccessfully to expand the mandate of his agency to include the regulation of pollution not only from the manufacture of chemicals, but also from the combustion of coal. To Fletcher, and those who shared his views, it made little sense to restrict sulphur dioxide emissions from chemical works while allowing ordinary factories free rein to emit sulphur-laden "black coal smoke," which he considered to be "the commonest and most prevalent of noxious vapours."[12] Although it would have been logical for the Alkali Inspectorate to regulate sulphur emitted during coal combustion, the idea faced insurmountable political hurdles. Most members of Parliament preferred to leave the administration of health matters in local hands, and municipal officials fiercely resisted all attempts to weaken their authority. Birmingham's health committee, for example, vigorously opposed any change that would have granted control over smoke inspection to the Alkali Inspectorate or to any other group it could not control.[13]

The Sanitary Act of 1866 marked a significant increase in parliamentary attention to coal smoke. For the first time local authorities throughout England and Wales were required (rather than simply permitted, as previously) to prosecute the owners of smoky factories. Although passage of the act was a significant event in the history of relations between central and local government, the law completely failed to reduce smoke. As with all national legislation against smoke until the Clean Air Act of 1956 (to be discussed later), the act put no restrictions on the amount of smoke that could be released from fireplaces and stoves in private dwellings. Unfortunately, some industries, such as iron and steel manufacturing, were explicitly exempted from the law on the grounds that it was impossible for them to conduct their operations without producing large amounts of black smoke.[14] In addition, a critical loophole existed in the wording of the statute, for it applied only to chimneys that released "black smoke in such quantity as to be a nuisance." Many defendants evaded fines by arguing that their furnaces produced smoke that was merely dark brown.[15] Others escaped punishment by claiming that they had taken every possible precaution to minimize the smoke they produced.[16]

It seems likely that the authors of the 1866 law chose their words in such a way as to allow all but the most egregious instances of smoke emissions to continue unchecked. Many people believed that considerable smoke was an inevitable concomitant of industry and that banning it would impose an unfair and undesirable restriction on manufacturers. Foreshadowing recent efforts by some in the United States to eliminate environmental regulations that they consider "burdensome" to industry, the framers of the Public Health Act of 1875 specified that black smoke was to be prevented only "as far as practicable."[17] Many factory owners won acquittals by arguing that their production processes unavoidably produced black smoke and that nothing short of going out of business would measurably reduce the amount of smoke they poured into the air. Municipal smoke inspectors proved quite sympathetic to this argument, and local officials exerted tight control over the hiring, firing, and everyday activities of the sanitary inspectors in their employ who had the legal authority to issue summonses to polluters. Some members of town councils were themselves owners of smoky factories or were close friends of such men. Not surprisingly, aggressive enforcement of the laws against smoke was often a low priority for councilors, and sanitary inspectors in such localities felt considerable pressure to be lenient with polluters.[18]

The Public Health (London) Act of 1891 took the regulation of smoke away from the police and gave it to the vestries (the local units of government that governed the parishes that made up London)—a change that gladdened those who disliked the growing influence of the London County Council, established three years earlier in 1888.[19] The 1891 act essentially extended to the metropolis, with the exception of the City, the provisions of the Public Health Act of 1875, which already applied to the rest of England and Wales.[20] This 1891 law included the declaration that smoke was illegal only if it was black, a loophole inserted at the insistence of the London Chamber of Commerce.[21] As with previous legislation, it failed to regulate smoke from dwelling houses, railroads, and government establishments such as the Woolwich arsenal.[22] The 1891 act substantially narrowed the range of what constituted illegal smoke and diminished the prospect of effective enforcement, but it also gave two important powers to the London County Council. First, if a vestry failed to take legal proceedings against a polluter in its parish, the council could intervene. Second, the council was empowered, albeit through a circuitous process involving the Local Government Board, to require vestries to hire additional sanitary inspectors.[23] The new law initially caused the smoke problem to increase in London. The main reason, as one critic noted, was that local officials were often "averse to convicting themselves or their friends; and a kindly and amiable transfer of good-will went on between our local governors, their friends, and inspectors. Inspectors did not see the smoke, and denied, therefore, that it existed—the end of their telescope was opaque."[24] By the middle of 1892 the London County Council had concluded that many local authorities were unwilling to enforce smoke laws. In response, it began collecting evidence of violations and forwarding them to local officials. Between 1892 and 1904 it issued over nine thousand notifications, and on eighteen occasions it declared that a local authority was in default of its responsibilities.[25] Despite these efforts, the ability of the London County Council to reduce smoke depended largely on the responsiveness of local officials and magistrates. In September 1898 a letter in the *Times* complained that judges treated polluters with excessive leniency. Pointing to a recent case in which the Metropolitan Electric Supply Company had been granted nine months to abate its smoke nuisance, the writer asserted that the only way to improve the air of London was to transfer enforcement authority to the London County Council.[26]

Birmingham provides another example of the limited effectiveness of local enforcement. The town contained a large number of specialized firms engaged in everything from metal rolling, wood turning, and the casting of brass to the manufacture of charcoal, nuts, bolts, nails, wire, buttons, and steel toys. Although the borough inspection committee levied numerous fines against the owners of smoky factory chimneys in the late 1860s, the fines were quite small, typically only ten or twenty shillings plus court

costs, and occasionally as little as a single shilling.[27] The situation sufficiently irritated the inhabitants of one neighborhood that in 1869 they sent
a petition to the town council complaining about dense smoke from several factories.[28]

Following the passage of the Public Health Act of 1872, Birmingham
instituted regular observations of factory chimneys to determine the duration and intensity of their smoke. During a two-week period in 1873 the
city's inspector discovered a wide disparity between firms. Cadbury's
cocoa factory on Bridge Street was among the best, with smoke visible for
an average of only a couple of minutes per hour, while Bolton's copper
works was much worse, with thirteen to forty minutes of smoke during a
similar period.[29] By the late 1870s Birmingham contained over a thousand
firms with registered steam engines and more than fifteen hundred industrial chimneys. Its smoke inspectors made over eleven thousand observations of chimneys in 1879 and reported that 389 firms had been observed
emitting dense smoke.[30] Birmingham followed a nonconfrontational approach toward polluters, emphasizing education and encouragement
rather than punitive measures. When a smoke inspector observed a factory emitting more than six minutes of continuous smoke, he would first
issue an informal caution.[31] Subsequent violations were reported to the
smoke subcommittee, which then sent a letter of warning to the owner.
Only if the factory continued to pollute would prosecution occur.[32]

During the early 1880s the number of firms reported to the health
committee for emitting dense smoke fell sharply, despite an increase in
the number of firms and chimneys in the borough. In 1881 and 1882 the
number reported was less than half what it had been in 1879.[33] Commenting on this decrease, the medical officer of health, Alfred Hill, claimed in
1882 that "as far as I can judge, there is a general improvement in the condition of the atmosphere."[34] One year later, however, the mayor chastised
the health committee and its inspectors for having ignored "the increase
of the smoke nuisance in the town."[35] The committee seems to have believed that its inspectors were adequately diligent, however, for it dismissed
the mayor's allegations by announcing that it saw no reason "to make any
new departure at the present time."[36] Yet the intervention of the mayor
appears to have either pressured or emboldened city smoke inspectors to
be more aggressive, for they reported nearly 50 percent more cases of dense
smoke to the committee in 1883 than in the previous year.

Six months after the mayor issued his complaint, the health committee fired chief nuisance inspector Thomas Hastings Dale, sending him away

with three months' pay "in lieu of notice."[37] The committee had long been dissatisfied with Dale. When he was hired in January 1881, the committee had offered him a starting annual salary of £250 and promised to increase it if he performed well.[38] In December 1882, nearly two years after Dale took office, the committee rejected a request by Dale for a raise at the same time that it boosted the salary of the veterinary superintendent by one-third and praised the latter's "ability and diligence."[39] Why did the committee display such animosity toward Dale? Was he too lenient in enforcing nuisance regulations, or overly forceful? The decline in smoke reports and prosecutions that took place during Dale's tenure may suggest that the committee viewed him as inefficient and wanted someone who would be more aggressive. It seems more likely, however, that Dale was dismissed because he sought to be more aggressive in enforcement than the health committee would allow. If committee members believed that Dale was doing too little to enforce smoke regulations, it would have been logical for them to use the mayor's complaints as ammunition against him. Their assertion that no changes were needed suggests that the committee rather than Dale was responsible for the problem of which the mayor complained. Further evidence that the health committee was not serious about the abatement of smoke and other nuisances can be deduced from the way in which it sought a replacement for Dale. Instead of conducting a national search—like the one it had begun two months earlier when it had placed advertisements for a new medical superintendent in the *British Medical Journal*, the *Lancet*, and the *Medical Times and Gazette*—the committee decided to seek a new sanitary inspector among its existing staff. In addition, it hoped to pay an annual salary of only £120, less than half of the £250 that Dale had earned and not even a third of what it paid the veterinary superintendent.[40] This suggests that the committee wanted to treat the appointment as a token position, one that would merely satisfy its obligation to have an inspector. The fact that local authorities in Birmingham and elsewhere possessed the power to make such decisions without any external review created significant potential for abuse.

The Limits of Local Regulation

Whether or not it happened in Birmingham, it is clear that in many localities, nuisance inspectors *did* experience pressure from local government officials. In 1880 Dr. Alfred Carpenter charged that many of "the members elected to serve on local boards are themselves offenders" against anti-smoke acts.[41] A few years later, the Manchester reformer Fred Scott asserted that municipal inspectors could not be trusted to "report offences created by persons upon whose goodwill they were virtually dependent for employment." As a corrective, reformers in Manchester hired their own smoke inspectors in the 1880s.[42] An 1886 survey of urban sanitary authorities in England and Wales reinforced the notion that manufacturers exerted a dominant influence on local health boards. In those towns and cities where authorities responded to the survey, 17.5 percent of the board members were manufacturers, a proportion second only to that of shopkeepers, who constituted 30.8 percent of the total. Presenting these findings at a conference of sanitary inspectors in York, the author of the study argued that the interests of manufacturers

> directly clash with sanitary work. Among their numbers are chemical manufacturers, tanners, bone-boilers, fellmongers, glue-makers, manure-makers, fat-refiners, brick-makers, dye-makers, candle-makers, soap-boilers, gas-makers, and others too numerous to mention, the nature of whose operations are such as to pollute our water and load the air of our populous places with noxious gases and smoke.

Such men, he concluded, were hardly the sort of people who ought to enforce health regulations. On the contrary, they were "eminently fitted to make formidable enemies to sanitary reform."[43] Two years after the survey, the Bolton industrialist Herbert Fletcher admitted that some local officials had a conflict of interest over the enforcement of smoke regulations because they were themselves the owners of polluting factories. But he remarked that personal gain was not the only factor that discouraged proper enforcement. Even conscientious officials, he explained, were reluctant to act against firms that threatened relocation to jurisdictions that were more tolerant of pollution.[44] Similar complaints came from the heavily polluted city of Sheffield, where the medical officer of health claimed that polluters escaped punishment because of friendly judges, not ineffectual smoke inspectors. Magistrates, he noted gingerly, were "often themselves personally interested in the subject" and consequently "out of sympathy with the efforts of the municipal officials charged with the carrying out of smoke prosecutions."[45]

During his tenure as medical officer of health for Salford in the 1870s and 1880s, John Tatham (1844–1924) faced strong opposition when he challenged "the right commonly assumed by manufacturers to pollute the atmosphere by smoke." At a meeting of the National Health Society in 1906 he charged that, with the exception of London, "medical officers of health throughout the country hold their appointments at the will of practically the smoke-producers themselves, who are all-powerful in their several neighbourhoods."[46] Mrs. Arthur Lyttelton, a former poor-law guardian in Manchester, offered an additional voice to these allegations, complaining that city officials "winked at the pollution of the atmosphere."[47]

There was little improvement in local enforcement by the end of the century. In 1899 the *Builder* complained that "mill-owners and the friends of mill-owners rule the local councils, and the nuisance-inspector, tempering knowledge with expediency, keeps his eyes on the nuisances at his feet and does not venture to look up lest he should see the nuisances above his head."[48] The best way to solve the smoke problem, the journal argued elsewhere, would be to transfer enforcement from local authorities to "county councils or some other body of men who would deal with offenders without fear or favour."[49] The same year that the *Builder* published this proposal (1900), the president of the Sanitary Inspectors' Association called on Parliament to remedy the "precarious" position facing local inspectors. He pointed out that although the Public Health Act of 1875 had prohibited local authorities from firing medical officers of health without the approval of the Local Government Board, it included no similar protection for sanitary inspectors. Another inspector, supporting the proposal, argued that "until we are made responsible only to some central authority, we cannot always do our duty."[50]

Prefiguring arguments today that strict pollution standards at the national level will drive jobs to countries with weaker environmental laws, Glasgow's chief sanitary inspector warned in 1900 that the adoption of more stringent antismoke regulations would harm the city's manufacturers and benefit competitors who operated in cities or towns with more lenient standards. To avoid such a prospect, he suggested that Parliament ought to establish uniform standards throughout Britain.[51] Similar calls came from Sheffield, where the city's chief smoke inspector warned that in the absence of centralized enforcement, polluting factory owners everywhere would continue threatening to leave whenever local authorities tried to make them reduce their smoke output.[52]

Another argument for central control was the fact that smoke did not respect jurisdictional boundaries but drifted wherever the wind carried it. As one writer observed in 1880, "The smoke of the Metropolis extends like a pall over a large tract of country," sometimes as far as eighty miles from London.[53] Rollo Russell pointed out that "the transparency of air over a great part of England would be noticeably greater if London insisted on the general adoption of reasonable and economical methods of cooking and warming," as opposed to inefficient and smoky coal fires.[54] Laws against smoke, however, defined the problem (and established a framework of regulation) on a local level. Polluting industries could therefore avoid punishment not by reducing the amount of smoke they released, but simply by erecting taller chimneys that would transfer the problem to more distant places. In 1881 Birmingham authorities dropped their prosecution of a firm that was violating smoke laws when they learned that it was increasing the height of its smokestack.[55]

Coal smoke long remained a low priority in Birmingham. Although some manufacturers adopted cleaner practices, many simply treated these modest fines as a cost of doing business and did nothing to reduce the clouds of smoke that they produced. Joshua Stubbs and Co. was an example of the latter; although the company had been fined five times in the previous two years for violating antismoke laws, in 1889 its neighbors complained about "the disgraceful nuisance caused by the continued emission of smoke" from its chimneys.[56] A dozen years later, Rev. Thomas Bass charged that the fines imposed on large firms for polluting the air were often smaller than the ones levied on petty bookmakers.[57] One of the places in his parish with the foulest air was named Oxygen Street. Struck by this unintentional irony, Bass declared that there was "so little oxygen in Oxygen Street that for a decade the death rate has been thirty-nine per thousand per annum"—nearly twice the national average. In addition to the usual sources of impure air, this street contained "an acid works, emitting blue clouds, which make the nose tingle, irritate the throat, and cause one to get their breath in gasps."[58] The rest of the parish contained additional sources of pollution:

> There is a stench from the gasworks at the bottom of Richard Street, and a pungent odour from the slimy canal. The chimneys of the works dotted above give stinks and smoke and smuts in abundance—I saw from Dartmouth Street in the middle of the day a dozen chimneys belching forth smoke as black as night—and other works send out an acid vapour that eats away the glass in the windows, ironwork of buildings, and even stone itself. It is at times so strong that it wellnigh stran-

gles the people, who, when they wake after a night's sleep, find the taste of acid upon their lips. The awful pollution of the air gives perpetual headache.[59]

Although many of the aerial contaminants that concerned Bass came from industrial processes, he also worried about the effect of bodily processes on the air. Like many housing reformers of the late nineteenth and early twentieth centuries, he reserved special condemnation for back-to-back houses and enclosed courts because they failed to allow cross-ventilation. He claimed that the air in some of Birmingham's housing courts was so foul that during visits to them he "had difficulty in staving off an attack of vomiting."[60] Bass's civic activism did little to endear him to municipal authorities, who often dismissed his pronouncements on public health matters as meddlesome intervention. In 1902, for example, the medical officer of health Alfred Hill commended the nuisance inspector for adhering to "medical direction" rather than submitting to "clerical dictation" from Bass.[61]

Enforcement of antismoke laws long remained weak in many other cities as well. An 1891 act for Glasgow established a minimum fine of £2 for a first conviction and £5 for a second.[62] In contrast to Birmingham, however, where the authorities did not prosecute unless more than fifteen minutes of black smoke came from a chimney within a single hour, Glasgow allowed only two to three minutes of black smoke.[63] Similar to the practice in Birmingham, polluting manufacturers usually received two warnings before prosecution was initiated. In 1900 401 "intimations of excess smoke" and 108 official warnings were issued. Only 36 cases were prosecuted, of which 33 resulted in conviction. The average fine was less than £1.[64]

Manchester too imposed inconsequential fines. Objecting to this, Rev. W. E. Edwards Rees complained that "there are people in Manchester who systematically pollute the air and systematically pay the fine—the ridiculous fine that is imposed . . . is a mere bagatelle." He asserted that many manufacturers found it "much cheaper to pay the fine" than to invest in cleaner technology.[65] Similar problems occurred in London. In 1899 the London County Council took several railroads to court in an attempt to force them to use clean-burning anthracite coal, but the threat of a £3 to £5 fine proved meaningless to companies that could save tens of thousands of pounds annually by using less costly—but smokier—bituminous coal.[66]

In Leeds enforcement was downright lackadaisical. Despite being one of the smokiest cities in Britain, barely a handful of firms were convicted of smoke violations in the first years of the twentieth century, and fines—when imposed—averaged just half a pound.[67] Eventually responding to criticism from the chemist Julius B. Cohen (1859–1935) and other members of the Leeds Smoke Abatement Society, the city obtained parliamentary approval in 1905 to toughen its smoke ordinance, hire an inspector with an exclusive responsibility for smoke, and raise the penalties for violators.[68]

Low fines were not the only hindrance to enforcement. In 1878 the owner of a brass foundry in Birmingham attempted to influence a smoke inspector by offering him two shillings (about half a day's pay). The inspector reported the incident to the health committee, and the latter addressed a sternly worded letter to the foundry owner, warning him that any further attempts at bribery would result in criminal prosecution. A short time later, the committee fined another brass founder for having attempted to bribe a smoke inspector by offering him a single shilling.[69] In many communities, however, no bribes were necessary to persuade local officials to look the other way. In 1896 the journal *Engineering* reported, "No magistrate would continue to fine a manufacturer for making smoke if its cessation entailed the closing of his works."[70] The excuse became codified in many industrial cities, where "protected trades" like steelmaking were exempted from smoke regulations. Leeds's chief nuisance inspector complained in 1900 that as a result of this exemption, "the worst offenders" were able to flout his authority.[71] Similar exemptions applied in Sheffield, with the acquiescence of its chief medical official. Although Sheffield's foundries emitted "dense volumes of the blackest smoke for periods not of minutes but of hours, and this at such a level that it quickly envelopes the neighbouring streets in a dense cloud," its medical officer of health viewed this pollution as a necessary evil. In his 1897 report on the health of the city, Dr. Harvey Littlejohn wrote that "as long as Sheffield trade consists in the manufacture of iron and steel and cognate processes, the atmosphere of Sheffield must necessarily suffer in comparison with that of other towns, and a certain amount of smoke and consequent discomfort is one of the penalties those who are connected with the trade of the town and who are compelled to reside in it will have to suffer." Suffer they did—in 1889 Edward Carpenter described Sheffield as a place where soot "jumped up and hit you in the face."[72]

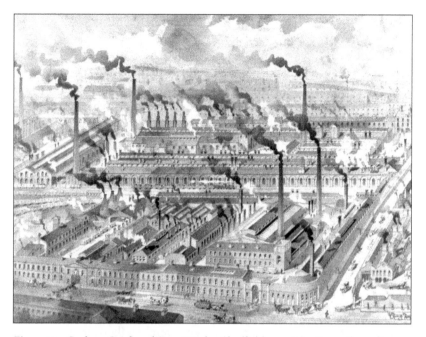

Figure 8.1. Cyclops Steel and Iron Works, Sheffield, ca. 1862. From the collections in Sheffield Local Studies Library (s09743); reproduced with kind permission.

Critics frequently complained that both the national government and municipal authorities set poor examples by failing to reduce the smoke produced from their own operations, such as arsenals, gasworks, and electrical power stations. In 1855 a member of the audience at a meeting of the Society of Arts asserted that "the government establishments at Woolwich and Greenwich were the largest smoke-makers in the kingdom, and that no attempts appeared to be made to abate the smoke nuisance there."[73] In Birmingham the Adderley Street gasworks, which was owned by the city, was cautioned in May 1880 for emitting ten minutes of dense smoke in a thirty-one-minute period.[74] Three years later, a city smoke inspector reported that he had observed the municipal gasworks on Windsor Street release thick smoke for fifteen minutes without interruption. No fines were issued, but an official caution was sent.[75] The health committee of the Birmingham city council heard complaints in September 1892 that the municipally owned Saltley gasworks was creating a smoke nuisance. The committee directed the nuisance inspector to look into the matter, but he later reported that the gasworks was not violating the law.[76]

This issue was later hotly contested in London, where groups such as the Liberty and Property Defence League and the London Chamber of Commerce decried city-owned businesses as "municipal socialism."[77] Although all of the gas companies in London were privately owned, many electrical power stations were owned by local authorities. In 1900 a representative of the London Chamber of Commerce complained to the *Times* that an electrical station owned by the vestry of Shoreditch was being allowed to produce smoke with impunity, while a privately owned station in Westminster had been fined £10 for an identical violation.[78] This problem finally began to be addressed in the early twentieth century when the General Powers Act of 1910 gave the London County Council the authority to levy fines against local authorities under its jurisdiction if they violated smoke restrictions.[79]

Measuring Pollution

Scientists and municipal inspectors used several methods to measure air quality, most of which relied on observations of the visible constituents of smoke. To estimate how much smoke a particular factory produced, inspectors observed individual chimneys and recorded the number of minutes that each emitted smoke. In February 1888, for example, smoke inspectors in Sheffield conducted 198 chimney observations of an hour each. These observations covered just a fraction of the hundreds of chimneys in the city that were attached to metallurgical furnaces and steam boilers.[80] Inspectors recorded 1,832 minutes of black smoke—an average of about nine minutes of smoke from each chimney per hour. Perhaps for political reasons, the report does not indicate how many chimneys exceeded the permissible level of ten minutes of dense smoke per hour, but it does state that only four notices were served on factory owners.[81] More detailed information comes from William Nicholson, who later recalled that when he became chief smoke inspector in about 1890, the air of Sheffield was among "the worst in the country. . . . Most of the boiler and furnace chimneys were emitting dense volumes of smoke and flame day and night. Many smoked continuously for hours, and the smoke was as black as the coal and as dense as could possibly come out of the chimneys."[82]

To counter criticism that their judgments were subjective and inconsistent, and to boost their professional standing, inspectors developed standardized scales to measure smoke. The most common of these was the Ringelmann chart, which contained shaded squares ranging from light gray to black. By matching the shade of smoke coming from a smokestack to the card that most closely resembled it, inspectors hoped to achieve an objectivity that would not vary from one observer to another or from one week to the next. Although this method obviated the use of imprecise descriptive words and may have given inspectors greater scientific legitimacy, it was far from infallible. As contemporaries noted, Ringelmann observations were affected by the time of day and the condition of the sky. Addressing the Coal Smoke Abatement Society in 1907, the German smoke expert Louis Ascher declared, "It is of great importance to find a good method of showing objectively the density and duration of smoke; subjective evidence is likely to be wrong. We have as yet no exact methods."[83] Even more problematic was the assumption—largely unquestioned at the time—that the health risks and environmentally harmful characteristics of smoke were proportional to how dark it appeared.

Although observing the smoke from a particular chimney was itself a challenging task, it was easy compared to the difficulties inherent in monitoring the quality of large bodies of air. Until the early twentieth century visual observations remained the dominant measure of ambient air quality. To quantify the amount of visible pollution, inspectors first measured the distance from a central observation point to landmarks at varying distances. By noting which ones were and were not visible at a given time, they were able to express air quality in terms of the number of feet or miles of visibility. Yet this method, like the Ringelmann chart technique, had shortcomings. A fundamental difficulty, particularly in a climate such as Britain's, was that of distinguishing the poor visibility caused by air pollution from that caused by naturally occurring fogs. This was compounded by the tendency of smoke to make existing fogs thicker and more persistent. It therefore became extremely difficult to determine which aspects of the weather were natural and which were caused by human actions.

As with observations of individual chimneys, measurements of air quality based on visibility ignored invisible combustion products, such as sulphur dioxide. Although contemporaries correctly identified sulphur dioxide as a damaging influence upon buildings, monuments, and plants, both reformers and the broader public paid greater attention to visible particulates. Many viewed smoke prevention as the more realistic goal, while sulphur dioxide emissions, prior to the development of effective control equipment, were usually accepted as a "necessary and unavoidable" product

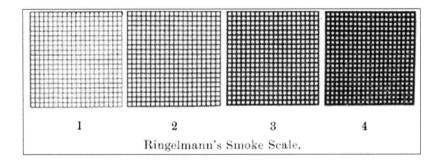

Ringelmann's Smoke Scale.

1   2   3   4

Figure 8.2. Municipal officials employed an idiosyncratic set of methods to measure smoke intensity. Pictured here are the Paris smoke scale and the Ringelmann smoke scale. From William Charles Popplewell, *The Prevention of Smoke, Combined with the Economical Combustion of Fuel* (1901), 127, 129.

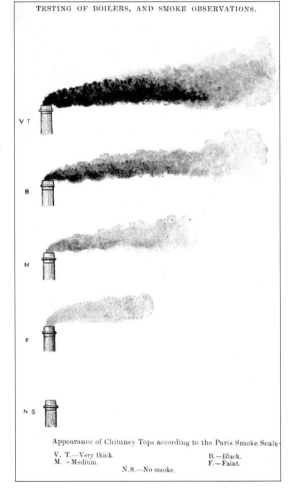

TESTING OF BOILERS, AND SMOKE OBSERVATIONS.

Appearance of Chimney Tops according to the Paris Smoke Scale

V. T.—Very thick.   B.—Black.
M. —Medium.   F.—Faint.
            N.S.—No smoke.

of burning coal.[84] More important, the concentration of sulphur dioxide in the air could be determined only through chemical analysis; visible smoke was easier to detect and harder to ignore. Thomas Coglan Horsfall suggested in the early 1890s that smoke prevention would indirectly reduce the damage from sulphur. If combustion occurred without forming smoke and the weather was dry, he claimed, sulphur would diffuse in the air rather than attach itself to soot particles and drift downward.[85]

To avoid some of the difficulties inherent in visual observations of the air, scientists developed techniques by which they could analyze the quantity and composition of soot and other particles that settled out of dry air or fell with rain or snow. Extrapolating from the small surface areas that they analyzed, researchers concluded that enormous quantities of soot and oily particles were accumulating in urban areas. In 1907 scientists at the University of Leeds measured the particulate matter in rainfall at ten stations in the city. They concluded that the amount of carbon, tar, and ash falling on Leeds ranged from 42 to 539 tons per square mile each year.[86] Two years later Glasgow's chief sanitary inspector, Peter Fyfe, estimated that 45 tons of "smuts" were falling to earth within the city each day, equivalent to an extremely high annual rate of 820 tons per square mile.[87]

During the early 1910s the Coal Smoke Abatement Society began a smoke monitoring program in London and encouraged antismoke groups and municipal health departments to set up similar programs elsewhere. By the mid-1930s, eighty-five local authorities and other bodies across Britain had established monitoring stations, which by this time were overseen by the Department of Scientific and Industrial Research.[88] Since a large proportion of a city's smoke drifted beyond its boundaries before its particulate matter fell to earth, such "smokefall" statistics merely hint at the magnitude of the air pollution problem, but they are useful for noting changes over time. In the early 1920s the average amount of soot (and road dust) that fell on Glasgow each year was estimated to be 326 tons per square mile, compared to 353 tons in London and 400 tons in Birmingham.[89] Glasgow's soot accumulation declined to 250 tons per square mile in 1938.[90] Large quantities of soot continued to fall on London even after the Clean Air Act was passed in 1956. In 1960 the Department of Scientific and Industrial Research reported that throughout Greater London at least 200 tons of carbon, ash, and dust fell per square mile, and that in some areas the amount was as high as 450 tons.[91] In places such as London, where coal was used mainly for heating, monitoring devices revealed that the levels of sulphur and smoke were four to five times higher in the winter than in the summer.[92]

Researchers also developed techniques to measure gaseous pollutants. During the 1870s Glasgow's sanitary department published detailed data on the concentration of carbon dioxide, sulphur compounds, and ammonia detected through chemical analysis of the city's air.[93] In 1880 the *Sanitary Record* praised this work, noting that Glasgow possessed "a series of stations at which the average condition of the air is continuously tested, night as well as day." According to the article, the only place where comparable measurements were performed was the Montsouris observatory in Paris.[94] In 1891 a committee of the Manchester Field-Naturalists and Archaelogists' Society published a study entitled *The Atmosphere of Manchester,* which contained data on the amount of sulphuric acid in that city's air. To aid its work, the chemist Julius B. Cohen devised a device that drew a known volume of air through a solution of hydrogen peroxide.[95] Using Cohen's device, the committee found that the concentration of sulphur compounds in the air was up to 100 times greater on foggy days than on clear ones.[96] A distinguished researcher, Cohen would later serve on the executive committee of the Smoke Abatement League of Great Britain and on the government's Committee on Smoke and Noxious Vapours Abatement. Perhaps because he had studied in Germany and still had close friends there, Cohen refused to help develop Britain's arsenal of chemical weapons during the First World War. He instead helped to devise gas masks and other defensive measures against them.[97]

Modern scholars generally share the view—widespread at the time—that the air of late-nineteenth-century Britain was exceptionally smoky. Eric Ashby and Mary Anderson note that the annual number of foggy days nearly tripled in London between the 1850s and the 1880s,[98] and Peter Brimblecombe estimates that smoke concentrations peaked in the metropolis around 1890.[99] Daily observations of fog at Greenwich between 1841 and 1950 suggest a similar conclusion: four of the five foggiest years during this 110-year period occurred between 1895 and 1905.[100] Data from the Meteorological Office station in Brixton in south London reveal a roughly similar pattern, in which fogs reached their greatest frequency in the late 1880s.[101] From both computer modeling and historical observations it is reasonable to conclude that London's air quality reached a nadir in the late nineteenth century, from which it recovered significantly around 1900. Part of this improvement can be attributed to the growing adoption of gas in place of coal. Yet much of the improvement probably resulted from a decline in the population density of London as people moved to the suburbs.[102]

During the last quarter of the nineteenth century local authorities in many of Britain's most polluted cities passed antismoke ordinances and hired smoke inspectors to enforce them. According to the annual reports that many cities issued, stricter laws led to a rapid improvement in air quality. Many observers disagreed and suggested that progress, if it had occurred at all, was minimal. As they pointed out, numerous hurdles hindered the abolition of coal smoke, including inadequate understanding of the technical factors involved, loopholes for certain industries, exemptions for smoke that was not black, a failure to legislate against household smoke, an absence of uniform regulations between different localities, a lack of consistent enforcement, and meager fines.

In the spring of 1914, after years of pressure by smoke abatement groups, the British government appointed a committee to consider whether additional legislation was needed to reduce the smoke problem. Despite an auspicious beginning, the outbreak of war that summer soon interrupted its work.[103] Because of concerns about aerial bombardment, coal smoke was actually encouraged during the war as a smokescreen for British cities. Yet, in other ways, the First World War may have stimulated smoke prevention.

The First World War ushered in an unprecedented level of government intervention in the economy of Britain. This, coupled with the expansion of government responsibilities connected with implementation of the "people's budget" shortly before the war, led many to look to government for solutions to problems that had long been seen as beyond its capacity to influence. Intervening in energy policy to reduce air pollution was such an issue.

As this occurred, scientists and government experts began to displace the authority that had once belonged to lay reformers. This development was not unique to the field of air pollution, but was instead part of a widespread professionalization of society. Although this trend began well before 1914, it accelerated considerably during the First World War. The Department of Scientific and Industrial Research, established during the war, took an early interest in air pollution as part of its efforts to promote efficiency. Its Fuel Research Board quickly emerged as a dominant sponsor of research into coal smoke, the results of which were disseminated in a series of technical papers.[104]

The governmental smoke investigation, known as the Newton Committee after its chair, reconvened soon after the war ended. Its interim report, issued in 1920, concluded that the bulk of Britain's smoke came from domestic rather than industrial sources, and it suggested requiring smokeless heating systems in all future government-sponsored housing projects. Reflecting a strong faith that technology would solve the smoke problem, the committee concluded by urging the government to "encourage the co-ordination and extension of research into domestic heating generally."[105] The publication of this report symbolized a new era in the response to coal smoke in Britain, one in which the terms of debate shifted from whether smoke was harmful to which technical means and public policy strategies should be adopted to prevent it.

According to one commentator, the great obstacle to cleaner air was the existence of what economists would later call externalities. As the Labour MP Percy Alden (1865–1944) explained in 1922,

> The real difficulty is that raw coal is cheaper to burn than it ought to be because the user does not have to pay for a great deal of damage which he does to other people. Similarly smokeless fuel is too dear. Suppose the State were to rectify this by taxing domestic raw coal to the extent of the indirect damage done, and by using the proceeds to subsidise smokeless fuels? There would then be no net taxation nor subsidy, and no harm done either to consumer or to miner, but a net gain would result to society because people would pay the real cost of what they bought, instead of buying cheaply something which does damage for which at present they do not have to pay.[106]

Despite the potential benefits of this proposal, it attracted little political support. Four years later, however, many antismoke activists celebrated the introduction of a bill in the House of Lords that called for raising the maximum penalty for smoke violations to £50 and which broadened the definition of smoke to include grit, dust, soot, and ash. To some, however, the bill was much too modest. As John Kershaw wrote in 1923, "It is always difficult to carry out legislation which is in advance of public opinion, and there is considerable doubt whether at the present time public opinion in this country is ripe for any very drastic legislation on the subject of smoke abatement." He suggested that instead of pushing for the passage of this bill, people who wanted cleaner air could better use their energies by focusing on "educational and voluntary methods" of reducing smoke. More work was needed, he believed, to convince the public that smoke was both an indication of wasteful combustion and a threat to health.[107]

Despite the bill's detractors, Parliament passed the proposed legislation, which became the Public Health (Smoke Abatement) Act of 1926. Unfortunately, the new law had little impact. As has happened on occasion in the United States, government agencies responsible for implementing environmental laws may find ways to write rules and implement policies that contradict the spirit of particular legislation. Although the 1926 act clearly stated that local authorities could apply for permission to regulate industrial smoke even if it were not black, such requests had to be approved by the Ministry of Health. According to its critics, the ministry rarely supported such requests. Twelve years later, on the eve of the Second World War, a legal expert with the National Smoke Abatement Society noted that "dense volumes of yellow, grey, green or brown smoke, however inimical to health, escape control under existing byelaws merely because such smoke is not black."[108]

CHAPTER 9

Pollution Displacement

In the next century electricity may undo whatever harm steam during the last century may have done . . . the future workman of Sheffield will, instead of breathing the necessarily impure air of crowded factories, find himself again on the hillside, but with electric energy laid on at his command.
—*W. E. Ayrton, 1879*[1]

If the cost of gas could be reduced by half it would in a few years do more towards solving the domestic smoke problem than all the reports ever written.
—**Engineering, 1922**[2]

BY THE LATE NINETEENTH century many people had come to believe that the production of large quantities of coal smoke was the inevitable price of "progress."[3] While the majority viewed this state of affairs as unfortunate, some, including John Ruskin, regarded it as proof that modern society was spiritually and environmentally bankrupt. The air would not be pure, they believed, until urban-industrial society was rejected. Others believed that environmental problems could be minimized without radical change. Technology, they suggested, could be harnessed to solve problems that technology had created. The *Builder,* though a strong opponent of smoke, also celebrated "how much of greatness, of energy, of healthy influence there has been in that era of machinery which he [Ruskin] despises."[4] Adopting the perspective articulated by the *Builder,* most reformers believed that it was unnecessary to reject modern technology to achieve clean air. The real problem, they insisted, was not modernity itself, but an unnecessary product of it: coal smoke.

Although some people considered the deleterious environmental effects of air pollution to be an indictment against technology, many others hoped that engineers would create a technical fix that would assist nature's ability to render pollution harmless. Admonishing his fellow engineers to give greater attention to air pollution, Douglas Galton complained in 1880 that in contrast to the extensive efforts that had been undertaken to improve London's air quality by removing garbage and constructing sewers, "very slight efforts have been made in the direction of purifying the air of towns from smoke."[5] Throughout the late nineteenth century scores of inventors designed supposedly smokeless fireplaces and steam boilers, none of which significantly reduced the production of air pollution. Other inventors, who observed that even gas and smokeless varieties of coal inevitably released some sulphur into the air, focused not on preventing smoke but on disposing of it. On many occasions from the 1850s through the 1880s the chemical manufacturer Peter Spence offered an ingenious proposal to solve the smoke problem. Expanding on the concept of sewers as conveyors of liquid and solid waste, he suggested that they could carry smoke from every fireplace in London to a thousand-foot-high smokestack, which would ensure that smoke was diffused into the upper atmosphere. His system, he boasted, would not only remove smoke but also help to neutralize and dispose of sewer gas.[6] An inventor from Wigan made a similar suggestion, proposing the construction of underground flues equipped with giant fans to carry smoke out of cities. Just as many sanitation experts had long hoped to convert sewage into a useful and profitable fertilizer, he predicted that valuable substances might be recovered from gases carried in the proposed smoke sewers.[7] In an attempt to evaluate the claims made by the manufacturers of such devices, the Sanitary Institute appointed a committee charged with testing them and providing the public with "some reliable means of judging their merits." Douglas Galton, who later played a prominent part in smoke abatement activities, was one of the leading members of this committee.[8]

Others, instead of looking for cleaner methods of burning coal, searched for alternative sources of heat and power. Not long after the invention of the incandescent light bulb, many hoped that electricity would soon provide people with a clean source of energy. Decrying the existing ways in which coal was mined and used, the *Electrician* complained in 1879 that "we degrade a large population to the most loathsome labour in the pits, we consume and poison the air, and we load it with such quantities of smoke that the modern sun is barely visible." Technology provided the means to reverse this process. Someday people would rely on electricity, generated by water power or clean coal-fired power stations. When

Figure 9.1. Observing infant electricity, Old King Steam asks Old King Coal, "What will he grow to?" Drawing by Sir John Tenniel in *Punch,* 25 June 1881, 295.

this day came, "A smokeless millennium will set in which will rejoice the heart of Mr Ruskin."[9]

Despite the predictions of some, electricity long remained too expensive to compete with coal for heating and cooking. In 1899 the *Builder* nonetheless held out the hope that electricity might eventually provide "the ultimate solution" to the smoke problem.[10] By this time, however, many people had grown skeptical about claims that electricity was a clean source of power. Coal-fired power stations often emitted large quantities of smoke, particularly during periods of heavy load. In the first decade of the twentieth century an electrical power plant in Greenwich, owned by the London County Council, sparked a vigorous debate about where such facilities would be sited. Because of its proximity to the Royal Observatory, critics complained that "smoke and acid vapours" from the works, as well as heat and vibration, threatened "to render the numerous valuable scientific observations made there useless."[11]

As electricity consumption grew in the decades that followed, its generation produced ever-larger quantities of smoke. In 1930 a committee of the London County Council complained that electrical power plants produced more smoke and grit than virtually any other source in London. One reason for this was the low quality of coal burned in generating stations. To save money, companies often used coal that had not been screened to remove dirt and other nonflammable matter.[12]

Although many continued to express hopes that electricity might some day be cheap enough to use for all purposes, most people expected to obtain heat from the direct combustion of fossil fuels for the foreseeable future. Three so-called "smokeless fuels" seemed promising as substitutes for smoky bituminous coal. Anthracite was a naturally occurring type of coal that contained much smaller quantities of volatile organic compounds than ordinary coal. Unfortunately, anthracite comprised only about a tenth of Britain's coal supply in the late nineteenth century and cost more than

Figure 9.2. If electricity could not abolish smoke, suggested this cartoon by Charles Harrison, perhaps it could provide a stronger source of light than the flaming torches used by the "link boys" who guided paying customers through thick fog. From *Punch,* 19 Nov. 1898, 229. Reproduced with permission of Punch Ltd.

bituminous coal. In addition, it was relatively difficult to ignite and produced, according to its critics, "a cold and gloomy fire."[13] Much more plentiful than anthracite were two other fuels—coke and coal gas—which could
be made from ordinary coal. Coke was already used widely as an industrial fuel, particularly in the production of iron and steel. Like anthracite,
it was difficult to ignite, which caused many people to view it as unsuit-
able for household use. In addition, neither coke nor anthracite produced
the colorful flame that many British householders expected from their
fires.

Although few visible signs of them now remain, virtually every town
and city in Britain once contained gasworks in which coal was transformed into gas, coke, and other by-products. During the nineteenth and
early twentieth centuries gas provided the main source of artificial light in
Britain, and coke served as an important industrial fuel. Electricity began
to supplant gas for lighting purposes in the late nineteenth century, but
consumption of gas and coke increased nonetheless as people adopted
them as relatively clean-burning alternatives to coal for cooking and heating. By the early 1960s—just before manufactured gas was replaced by
natural gas from the North Sea—gasworks in Britain were consuming 22
million tons of coal each year, and coke plants were using even more.[14]

Despite the fact that gas and coke were smokeless at the point of consumption, their production was an exceedingly dirty process that damaged the health of workers and nearby inhabitants and poisoned the air,
soil, and water with a host of hazardous by-products. Gas and coke did
not eliminate pollution; instead, they displaced it from one environment
and group of people to another. In addition to redistributing the burden
of pollution spatially and socially, the manufacture of gas and coke also
displaced pollution chronologically by bestowing a toxic legacy on future
generations.[15]

Historians have devoted little attention to the environmental consequences of the gas industry in Britain. Although several studies provide
valuable insights into its technological, organizational, and financial aspects, they say practically nothing about its effects on workers or the environment. Such an oversight is not unique to this industry or to studies
of Britain. As Christine Meisner Rosen and Christopher C. Sellers recently
observed, scholarship in economic and business history has "tended to
treat industrial impacts like pollution as well as most other environmental dimensions of business activity as if they were what economists call
'externalities.'" Jeffrey K. Stine and Joel A. Tarr have similarly noted that

historians "have on the whole neglected not only worker safety but also the environmental consequences of industry and manufacturing."[16] Historians of air pollution in Britain have displayed a similar propensity to ignore pollution from gasworks and coke plants. Focusing on the serious problem of coal smoke, they have considered how the consumption of gas and coke affected the atmosphere and have overlooked the environmental consequences of their production. Eric Ashby and Mary Anderson, for example, assert that "London air became cleaner in the first decade of the twentieth century" largely as a result of "the enterprise of the gas industry." Although air quality improved in many places as a result of gas and coke, the production of these fuels also created new hazards to human health and the environment.[17]

Gas and Coke Production

The first public gasworks in Britain began operating in 1813. Owned by the Gas Light and Coke Company, it was located in Great Peter Street in Westminster. Gas quickly gained popularity as a source of light in streets, commercial buildings, and private houses; companies soon began producing it throughout Britain. By the middle of the nineteenth century, 760 towns in Britain and Ireland possessed at least one gasworks. All over the kingdom workers dug up streets to lay pipes for gas—London alone contained 2,000 miles of mains by 1850. The amount of coal used by the gas industry rose from just 500,000 tons in 1830 to 10 million tons in 1887. London had by far the largest gas production in Britain throughout the era of manufactured gas. At the beginning of the twentieth century its gas companies consumed approximately 4 million tons of coal each year.[18]

Both coke and gas were created by the "carbonization" of coal in retorts that were heated to approximately 1,300°C. The bituminous coal that was often used to heat these ovens produced a great deal of smoke, and fugitive emissions from coal undergoing carbonization further contaminated the air with a mixture of smoke and foul-smelling vapors. Despite becoming red-hot, the coal did not burn, because the retorts were tightly sealed to keep out oxygen. Instead, the extreme temperatures forced virtually all of the volatile constituents from the coal. In some places, particularly during the nineteenth century, the sole purpose of carbonizing coal was to make coke. The "beehive" coke ovens that dotted Britain's industrial landscape made no attempt to capture the gases and liquids that were driven from coal as it was heated. Parts of northern England became virtually denuded of vegetation as a result of this pollution. In contrast to beehive ovens, "by-product" coke plants and gasworks produced a wide range of materials in addition to coke.[19]

Gasworkers, in an effort to secure not only better wages and shorter hours but also improved health and safety standards, formed a labor union in 1889. Will Thorne (1857–1946), who organized and led the National Union of Gasworkers and General Labourers and eventually became a Labour MP, possessed an intimate knowledge of the conditions in gasification plants. When he had worked at the Saltley gasworks in Birmingham, Thorne's job had required him to discharge coke from the ovens. Each time an oven door was opened, a puff of hot gases would burst out and explode into flame as it mixed with oxygen in the air. Describing conditions in the retort house, he recalled, "The work was hot and very hard. As the coke was drawn from the retort on to the ground, we threw pails of water on it, and the heat, both from the ovens and the clouds of steam that would rise from the drenched coke, was terrific."[20]

Workers in coke plants and gasworks faced an ever-present risk of injury and death. Records from the South Metropolitan Gas Company reveal the range of injuries to which workers fell victim. Between 1892 and 1896 ten of its employees were killed at work. Between November 1896 and December 1897 thirty workers at the company's East Greenwich complex missed work because of accidents on the job. These injuries included a severed leg ("taken off by coal crusher"), a lacerated face from being "pitched headlong from retort charger," mangled fingers and toes, a fractured skull, and numerous burns. Records from the Sheffield United Gas Company tell a similar story. During the late nineteenth century, a serious injury occurred at its Neepsend works approximately once a month.[21]

In addition to the risk of being maimed or killed, gas and coke workers experienced high rates of chronic health problems. A prominent medical expert claimed in 1930 that over half of the "notifiable cases of cancer" in Britain resulted from exposure to by-products created by the carbonization of coal in gasworks and coke ovens. One of the most hazardous of these by-products was tar. Despite using goggles and scarves to limit their exposure to it, gasworkers frequently suffered from blisters, boils, and warts. Later research revealed that those whose work brought them into close contact with the products of coal carbonization for over five years suffered from ten times the normal incidence of lung cancer. Writing in 1930 to the head of the Trades Union Congress, a worker in South Wales noted,

Figure 9.3. Photograph of coke being removed from the retorts at the Windsor Street Gasworks, owned by the city of Birmingham, ca. 1900. Reproduced courtesy of Birmingham Library Services, Local Studies and History (Wk/52/63).

> It was with much pleasure that I read in the *Daily Herald* . . . that you intend interviewing the "Home Secretary" re coke & bye product workers' industrial deceases [*sic*]. A subject in my opinion long overdue. I feel I can claim some knowledge of the terrible manner in which men engaged in the above industry are afflicted. I have worked and passed through some of the worst processes to be found on a bye-product plant for nearly 13 years; and I can assure you that I have suffered terribly and in consequence have lost a considerable amount of time.

He ended by noting that he had been out of work for five months and was "feeling all the better for it."[22]

Although conditions inside gasworks and coke plants were harsh, unhealthy, and sometimes deadly, an 1878 handbook on gas manufacturing claimed that gasworks provided "ample arrangements for the comfort of the stokers." These reassuring words were belied, however, by warnings that the sulphuric acid fumes that filled the air of retort houses would eat through unprotected iron. Expressing greater concern for damage to property than to people, the book noted that even galvanized nails had to be coated with tar to prevent their being "rapidly destroyed by the action of the gases and vapours necessarily present in buildings of this description."[23]

Despite growing evidence that gasworkers experienced disproportionate levels of certain diseases, the industry long continued to deny that coal gasification posed any dangers to health. The *Gas Bulletin,* a publication of the British Commercial Gas Association, asserted in 1933 that "people employed in gas works never suffer from headaches, because of the fact that ammonia, a by-product of coal carbonisation, is a cure for this complaint." Physicians who examined workers faced a potential conflict of interest because such visits were often arranged and paid for by the employer. Practitioners who made diagnoses that companies disliked were unlikely to receive future referrals. In the case of one South London gas company, a single physician examined numerous employees over a period of many years; in virtually every case, he declared that their illnesses had no connection to their working conditions. When an employee complained in 1947 that working inside the purifier box was making him ill, the chief engineer sent him to this doctor with a note that asked for a report on his "affectation." Upon examining the man, the physician concluded that the "pain and vomiting has nothing to do with his work." After it nationalized the industry in 1949, the government began conducting regular medical examinations of gasworkers in an effort to detect health problems before they reached an advanced stage. Thirty employees of the same gasworks were examined in 1954; five of them were diagnosed with cases of eczema, warts, and melanoma.[24]

Pollution

In addition to posing considerable hazards to their workers, gasworks filled their neighborhoods with smoke and foul smells. During the 1820s individuals who lived near the Gas Light and Coke Company's works in Westminster complained that its fumes harmed plants and trees, sullied clothing, tarnished brass and copper, discolored paint, and impaired their health. Those who were able to move away from the immediate vicinity of this and other gasworks tended to leave. Over time, such localities became occupied almost exclusively by poor and working-class people, who rarely possessed sufficient economic, political, or legal power to challenge polluters' activities. As a London newspaper saw it in 1864, "Wherever a gas-factory—and there are many such—is situated within the metropolis, there is established a centre whence radiates a whole neighbourhood of squalor, poverty, and disease." Nearly a century after its beginning the gas industry continued to produce large amounts of air pollution. As the *Lancet* observed in 1904, each time retort doors were opened to remove coke and add unprocessed coal, "volumes of thick black smoke" poured into the air.[25]

In contrast to coal combustion, the primary environmental consequence of which was air pollution, the carbonization of coal directly polluted soil and water as well. As an engineer employed by the Gas Light and Coke Company noted in 1907, many by-products "long remained in the category of 'impurities' which had to be got rid of somehow." Derivatives that appeared to have no commercial value were often allowed to simply drain into the nearest stream or river. Evidence of water pollution from gasworks can be found as early as 1821, when fish and eels in the Thames were reportedly killed as a result. Although the Gasworks Clauses Act of 1847 barred the industry from discharging liquid effluents directly into watercourses, its impact was limited. Many companies dealt with such wastes by pouring them into holding ponds, from which they could seep into groundwater or "accidentally" overflow into nearby streams or rivers. Solid wastes were often used to fill in low-lying areas, polluting not only the soil but frequently the air and water as well. During the 1930s, six decades after gasification ceased there, 6,000 gallons of coal tar were removed from the soil beneath the former works of the Gas Light and Coke Company in Westminster. Fittingly, this site became the headquarters in 1971 for the Department of the Environment. Eventually, many wastes were reused instead of being buried or poured away. For example, the slaked lime used to absorb impurities from gas was often sold as an agricultural fertilizer. Yet the reuse of waste materials from gasworks and coke plants often meant that toxic substances simply entered the environment via different pathways.[26]

Aware that its activities were degrading the environment, the gas industry worked hard to shield itself from lawsuits. In 1843, the owners of a brewery located at Earl Street and Horseferry Road in Westminster complained that their wells had become contaminated "by the draining and passing of water and liquor of a deleterious nature from the . . . Gas Light and Coke Company in Peter Street." As a result, the brewery could no longer use its own well water to make beer. The gas company, anxious to avoid a civil suit, paid the brewers £500 compensation in exchange for an indemnity against all past and future damage. Three years later, the company paid £790 to the owners of another neighboring property in a similar agreement. Although these sums were substantial, lawsuits could prove even more expensive. Such a case arose in Scotland during the first years of the twentieth century when the brewing firm James Muir and Son sued the Edinburgh and Leith Gas Commissioners for £10,000 in damages. The judge in the case ruled that pollution from the city-owned gasworks had contaminated the claimant's well "to such an extent as to make it absolutely unfit for brewing purposes." As compensation, he ordered the gas company to pay the brewery £3,730 plus legal expenses.[27]

People who lived near gasworks faced not only contaminated air and water, but also the risk of injury in the event of an accident. Leaking gas could cause suffocation or explosion, and accounts of such events appeared with disconcerting frequency in newspapers and magazines. A large explosion at the Nine Elms Gasworks in the London borough of Battersea killed nine people in 1865. According to a contemporary account, "People nearly a mile off were thrown violently down, and persons who were in houses and streets adjacent to the works received severe burns." Shortly after this accident, an anonymous letter to the *Times* urged the removal of gasworks from densely populated districts. The writer argued that this would not only limit casualties in the event of an explosion, but would also reduce the number of people who had to endure the habitual stench of gas production. For despite "every precaution it is impossible to produce gas on a large scale without contaminating the surrounding atmosphere with offensive, if not noxious effluvia. Gasworks are a positive and unmistakable nuisance to all whose olfactory nerves are not smitten with paralysis."[28]

The environmental impact of individual gasworks increased during the economic boom of the 1850s and 1860s, spurred not only by rising demand, but also by technological developments that allowed gas companies to achieve large economies of scale in both production and distribution. In 1867, two years after the explosion at Nine Elms, the Gas Light and Coke Company purchased a marshy site along the north bank of the Thames in East Ham, where it built the mammoth Beckton gasworks, once the largest in the world. This installation consumed vast amounts of coal, and the gas it produced flowed to central London through gas mains that were four feet in diameter. Another huge gasworks in London was the South Metropolitan Gas Company's complex in East Greenwich (today the site of the Millennium Dome). In the early years of the twentieth century

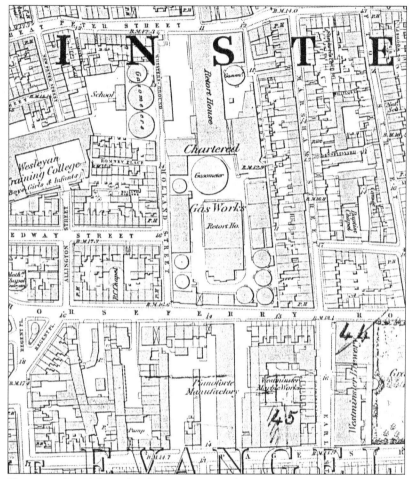

Figure 9.4. Gas Light and Coke Company Works in Westminster, 1869. Note its proximity to houses, schools, and a brewery. Detail from the 1869 Ordnance Survey Map of London.

it consumed 2,200 tons of coal each day and produced 23 million cubic feet of gas. Alarmed by the company's proposal to expand it, a nearby resident complained in 1902 that unless "some new process by which gas can be produced without filling the atmosphere with dust and dirt" were discovered, "the beneficent influences of Blackheath and Greenwich Park will be counteracted."[29]

As a result of these changes, the pollution that occurred during carbonization became increasingly concentrated in areas where land was cheap

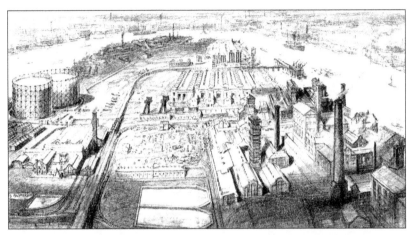

Figure 9.5. East Greenwich Gasworks, ca. 1924. From *A Century of Gas in South London* (1924), facing 20. Reproduced courtesy of National Grid Transco, plc.

and residents lacked influence. The experience of living in such a place made a deep impression on Robert Roberts, growing up in Salford in the early twentieth century. Although his neighborhood contained several sources of air pollution, including a brickworks and an iron foundry, he considered the local gasworks—which filled the air with thick smoke and a terrible stench—to be the worst offender.

On the other side of the class divide, the scientist Walter Hepworth Collins suggested that poor people did not object to living near gasworks because they possessed crude sensibilities. Addressing a meeting of the Sanitary Institute in 1890, he asserted that gasworks were typically "situated at the lowest accessible level, and, particularly in the case of old works, are surrounded by cottage or other property of an indifferent character; the adjacent neighbourhood being tenanted usually by the lower labouring class, whose sense of smell would not appear to be of a cultivated or refined type." Indignant in the face of such attitudes, Roberts wrote, "Our own streets stood immediately under the gasworks in the path of prevailing winds. Sometimes the air stank abominably for days on end. But very few questioned the right of industry to ruin our health and environment; in pursuit of profit the poor were expendable."[30]

Although local authorities often turned a blind eye to pollution from gasworks and other industries, this was not always the case. In 1899 the Garston Urban District Council notified the Liverpool United Gas Light Company that its operations were causing "pollution of the atmosphere by the escape of offensive odours." Frustrated by the company's failure to correct the problem, the council took it to court in 1901. Testifying at the trial, one resident complained that as a result of the gasworks, his family "had suffered from acute headaches, pains in the stomach, and abdomen, causing sickness and nausea and sometimes choleric pains. He had had to shut his windows to keep out the smells." Another testified that the smell had once so permeated his house that he had "vomited nearly all night." He concluded by complaining that "the neighbourhood was not really fit for anyone to live in."[31]

Interestingly, the main source of the council's complaint was not the smoke and vapors produced during carbonization, but the smells that entered the air as sulphur compounds and other impurities were removed from gas. Although manufactured gas was cleaner than coal, it was far from pristine. Gaslights imparted a sharp odor to the air of rooms in which they were used, and many people complained that they injured houseplants and deposited a sooty residue on ceilings. Legislation gradually forced gas

Figure 9.6. Gas scrubbers, Nine Elms Gasworks, London, ca. 1870. Note the pile of lime in the open shed. From Thomas Newbigging and W. T. Fewtrell, eds., *King's Treatise on the Science and Practice of the Manufacture and Distribution of Coal Gas,* vol. 1 (1878).

companies to reduce the level of impurities in the gas they sold, and many companies—hoping to encourage greater consumption—sought to achieve a higher standard of purity than the law required. Yet cleaner gas for the consumer often led to dirtier conditions at and near gasworks. Purification did not eliminate contaminants from gas; it simply concentrated them. According to the allegations brought by the Garston council, the gas company's scrubbers produced a stream of liquid with "an abominable stench," as did the iron oxide and slaked lime that were used in another stage of the purification process. The solicitor representing the company admitted that "gas works did give off effluvia" and "disagreeable smells," but he argued that they caused no harm. Going further, he put on the stand a consulting engineer who claimed that the smell complained of was actually beneficial, since it was "practically that of napthaline [*sic*], which was a powerful antiseptic and disinfectant." Recognizing that it would be foolhardy to rely solely on such a claim, he argued that the crucial question was not whether the firm had caused damage, but whether they had "done what they reasonably could to prevent the business injuring the neighbourhood." The magistrates who heard the case failed to reach a consensus, and the company was declared not guilty.[32]

Although its representatives frequently claimed to be doing everything possible to minimize its detrimental effects, the gas industry was slow to invest in cleaner technology unless doing so would directly benefit its bottom line. Two innovations began to transform the process of coal carbonization in the years around 1900. The first of these was the development of automated methods of stoking retorts with coal and discharging coke after carbonization. In an 1894 report to the Birmingham City Council, the committee in charge of the municipally owned gasworks claimed that mechanization would increase the amount of gas produced, improve its quality, lessen workers' exposure to unhealthy and dangerous conditions, and, most importantly, save money. Interestingly, the committee's report failed to mention another often-touted advantage of mechanical stoking: that it would reduce the emission of smoke. To raise this issue, the committee would have had to admit that its gasworks produced smoke, something it was extremely reluctant to do. A second major advance was the development of continuously operating vertical retorts. In contrast to the long-standing practice of roasting coal horizontally in small batches, continuous carbonization was less labor intensive, more energy efficient, and produced less smoke. Despite the economic and environmental advantages of vertical retorts, the cost of replacing existing equipment discouraged gas com-

Figure 9.7. Smoke from the Great Central Gasworks drifts over houses in Poplar, London, 1924. Reproduced courtesy of Corporation of London, London Metropolitan Archives (65.3 POP, negative 92/29).

panies from doing so. In the late 1940s over a third of the coal used in British gasworks continued to be processed in horizontal retorts.[33]

Consumption

Prior to the 1880s virtually all of the gas produced in Britain was used for lighting. Although a variety of gas appliances became available from the middle of the nineteenth century, the public generally regarded cooking and heating with gas to be expensive, ineffective, and possibly dangerous. Few gas companies attempted to dispel this view, confident that they could simply concentrate on supplying gas for lighting purposes. One exception to this occurred in London, where the Crystal Palace District Gas Company began promoting gas as a heating and cooking fuel in 1869. Besides sending each of its customers a pamphlet touting the benefits of gas, it offered to rent out heating and cooking stoves to them.[34]

The gas industry's complacency disappeared abruptly with the emergence of competition from electricity in the late 1870s. Eager "to keep up with the spirit of the age," several London retailers began experimenting with electric lights in their shops. Not to be left behind, the House of Commons followed suit in 1881. Amid predictions that electricity would soon make gas obsolete, the German-born electrical engineer and entrepreneur Sir Charles William Siemens (1823–83) argued that the gas industry could not survive if it continued to define itself primarily as a purveyor of lighting. "If gas companies and corporations rightly understand their mission," he declared, "they will take timely steps to supply . . . heating gas at a greatly reduced cost, the demand for which would soon be tenfold the gas consumption of the present day."[35] The promotion of gas for cooking was even more attractive to many in the industry, for it could help to equalize demand throughout the day and the year. In Leeds, for example, the municipal gasworks operated at only one-fourth of its peak capacity during the six warmest months of the year.[36] Cooking was largely unaffected by seasonal variation, and it took place mostly during the hours of daylight, which helped to counterbalance the reduced use of gaslight when the sun was up.

Although electricity eventually did replace gas as a source of lighting, the latter's demise was not nearly as rapid as Siemens had feared. The reason that gas retained its dominance as a source of lighting until the early years of the twentieth century can partly be explained in terms of technological inertia: considerable investments in fixed capital were required to convert from gas to electric lighting. Just as significant, however, were technological changes to gaslights themselves. For most of the nineteenth century gas customers relied on crude lamps that burned with an open flame. Although appliances of this variety were easy to build and required little maintenance, they possessed several disadvantages. First, the light produced was quite dim. Second, their brightness depended on the gas having a high luminosity when burned. Yet some types of gas burn with a nearly invisible flame, despite producing great heat. Gas companies had to choose between maximizing the luminosity or the heating value of the gas they sold. As long as government regulations specified that gas had to have a given lighting power, it made little economic sense for companies to promote gas as a fuel for heating and cooking. Carl Auer von Welsbach's invention of the gas mantle in 1886 revolutionized gas lighting. In contrast to traditional gas lamps, which burned gas in the open air and relied on the illuminating power of gas itself, the mantle glowed brightly when heated by burning gas.[37] This design obviated the need for gas that burned with a luminous flame. In a speech to the Smoke Abatement Conference of 1905, the head of a large London gas company called for new regulations that would allow the industry to reduce the illuminating power of the gas it supplied so that more heat could be produced from it.[38] This proposal became a reality with the passage of the Gas Regulation Act of 1920, but by the time it was enacted many people had already switched from gas to electric lighting.[39]

While Siemens based his arguments primarily on what would best ensure the future of the gas industry, he maintained that what was good for the industry was also good for the country. The use of gas for heating and cooking, he suggested, would result in "a radical cure of that great bugbear of our winter existence, a smoky atmosphere."[40] Many clean air advocates shared Siemens's view that the best way to reduce the smoke problem was to substitute gas for coal. Ernest Hart, prominent public health reformer and editor of the *British Medical Journal,* declared in 1883 that "the ultimate line of progress" lay in the general adoption of gas ovens and fireplaces. Supporting this approach at a meeting of the Coal Smoke Abatement Society in 1901, Sir Oliver Lodge (1851–1940) argued that "cheap gas . . . for cooking and heating," not "some new scientific discovery," would solve the smoke problem.[41]

A further advantage of gas, argued some individuals, was that it allowed consumers to avoid contact not only with coal and smoke, but also with the allegedly dirty and ill-mannered workers who were associated with the use of coal. Until only a few decades ago, chimney sweeps formed a large and immediately recognizable part of the British workforce. Notwithstanding the vital role that they played in maintaining the technological systems on which urban life depended, chimney sweeps, like sewer men and trash collectors, were often looked down upon by those they served. Smokeless heating and cooking allowed middle- and upper-class householders to alter the social as well as the atmospheric character of their immediate environment. As one member of the Coal Smoke Abatement Society explained in 1903, "with gas fires the noisy, dirty, expensive, and alarming sweep disappears, as indeed he does with coke and anthracite fires."[42]

Resistance

Although few liked the smoke that resulted from burning bituminous coal, many consumers nonetheless hesitated to forsake its use. In the minds of most people, coal possessed advantages over gas in terms of economy, convenience, sentiment, and even health. To start with, gas was comparatively expensive. Even Thomas Fletcher, who owned a factory that produced gas appliances, admitted in 1881 that it cost ten times more to heat a room with gas than with coal.[43] The price of gas decreased in subsequent years, but gas heating continued to be more expensive than coal. In 1886 Douglas Galton estimated that gas cost four times more than coal in terms of heating power.[44]

Aside from considerations of economy and health, there was a deep-seated psychological attachment to the traditional open fireplace. The *Lancet,* which strongly supported smoke abatement, nonetheless declared in 1892 that

> [w]e must look for a solution of the smoke and fog difficulty in the universal adoption of a grate that will burn the coal perfectly and economically, and yet be so constructed as to meet perhaps one of the most formidable of obstacles—namely, that of sentiment. It is hardly to be anticipated that a closed stove or a stove heated with gas or even electricity, however perfect its working may be, will ever replace the homely fire of glowing coals, which in the English mind will always lend a charm to the cold and dreary days and nights of winter.[45]

In addition to economic and cultural arguments against gas, some asserted that it was unhealthy. Aside from the risk of explosion, many believed that the everyday use of gas was harmful because it imparted a burnt odor to the air and made it too dry. In contrast to this unpleasantness, *Nature* credited open coal fires with promoting "freshness of complexion" and good eyesight.[46] Others criticized gas fires for failing to generate an adequate draft up the chimney. As a letter to the *Lancet* noted in 1892, open fireplaces, notwithstanding the smoke that they produced, were an excellent way of ensuring that rooms were adequately ventilated.[47] Lumping these concerns together with a tirade against universal education, the writer of a 1906 letter to the *Lancet* complained, "Gas cookers are not an unmixed good. I believe the penny-in-the-slot stove has almost as much to do with the appalling infant mortality among the poorer classes as the compulsory sending of all the *helpful* children to school. Clothes are not aired; the room is not ventilated by fire; there is no inducement for the mother to sit down and let the baby stretch its little limbs in the warmth."[48] Although many praised the open fireplace for its powers of ventilation, critics complained that this resulted in a significant loss of heat. Thomas

Coglan Horsfall decried the waste it entailed, arguing that "if our object were to lose as much as possible of the heat given by our fuel, and to make the little obtained affect as small a part of our rooms as possible, we could not find a system much better adapted for gaining our foolish purpose than the system by which we now try to warm our rooms."[49]

Promotion

During the second half of the nineteenth century many large towns and cities acquired ownership of gasworks and other utilities. Birmingham, Glasgow, Leeds, and Manchester all had publicly owned gasworks by the late nineteenth century, although in London and Sheffield gas was still supplied by private firms. Advocates of municipal enterprise put forth two reasons for it. First, although most of them supported laissez-faire principles, they believed that monopolies constituted a case of market failure. They argued that given the absence of competition, government intervention was necessary to ensure that gas was supplied to consumers at fair prices and acceptable quality. Second, many asserted that gas was too important to be controlled by private monopolies whose driving concern was profit maximization. Without the need to generate profits, gas could in theory be supplied more cheaply to consumers. Alternatively, profits could be generated as before, but used to "relieve the rates" (reduce taxes). Many municipalities adopted the latter approach, which disproportionately benefited large property owners. These earnings could be substantial; between 1844 and 1887, Manchester's gas department contributed over £1.3 million to city coffers.[50]

Municipal ownership of utilities affected not only the economic environment of towns and cities, but also the natural environment. Local authorities were responsible for establishing and enforcing environmental regulations for gasworks and all other industries not under the supervision of the Alkali Inspectorate. Where a local authority profited from an industry that it was supposed to regulate, a clear conflict of interest existed. In Manchester, one resident asserted in 1879 that the municipally owned gasworks was the city's largest single source of smoke. Although free-market advocates used such cases to attack what they decried as municipal socialism, private ownership was no guarantee that gasworks would face rigorous scrutiny. City councils and the sanitary committees they appointed were often dominated by businessmen who shielded themselves and their friends from unwelcome intervention.[51]

Although smoke abatement activists did much to promote gas, they frequently criticized the industry over the issue of pricing. In the 1880s reformers in Manchester urged the city to lower the price it charged. Despite entrenched resistance to their proposal, they continued to push for cheaper gas. The *Lancet*'s Manchester correspondent predicted in 1894 that the city's air quality would improve dramatically if gas replaced coal as the fuel for heating, cooking, and engine power: "But the gas committee stands in the way. The price of gas is too high for these purposes, and the committee is reluctant to lower it as the profits are used in aid of the rates." In 1912 the local branch of the Smoke Abatement League of Great Britain published a pamphlet that sought to turn the city's justifications for expensive gas upside-down. The booklet claimed that "if the ratepayers would forego the profit of £50,000 a year which they make on gas . . . they would be going a long way towards doing away with the loss of £700,000 a year which is caused by smoke."[52]

Beginning in the 1880s the gas industry gave substantial financial support to smoke abatement groups and collaborated with them in sponsoring exhibitions and conferences that promoted both their product and clean air. In 1883 the Gas Light and Coke Company and the South Metropolitan Gas Company each donated £100 to the National Smoke Abatement Institution. The relationship between the gas industry and the smoke abatement movement grew even closer in subsequent decades, thanks in large part to Sir George Livesey (1834–1908), director of the South Metropolitan Gas Company. In 1905—nearly a century before his ghost was reported to be haunting the Millennium Dome, built on land where his large gasworks had long stood[53]—Livesey chaired an air pollution conference sponsored by the Coal Smoke Abatement Society. The promotional arm of the industry, the British Commercial Gas Association, forged close ties with the smoke abatement movement. Its annual conferences included a section that highlighted the role of gas in clearing the air of smoke from coal fires, and the gas industry continued to give money to antismoke organizations. In return, smoke abatement groups welcomed advertisements that promoted gas and other "smokeless" technologies.[54]

These efforts paid handsome returns. The amount of gas sold in Britain tripled in the three decades preceding World War I as more consumers adopted it for cooking and, to a limited extent, heating as well. Gas cooking began to reach working-class households in the 1890s, following the invention of a coin-operated meter that allowed customers to pay for gas as they needed it. In 1898 a representative of the gas industry declared that 150,000 "penny-in-the-slot meters" were being used in London and claimed that the Gas Light and Coke Company had leased so many gas stoves that if placed side-by-side they "would reach from Charing-cross to the West-pier at Brighton and 50 miles back again."[55]

Cooperation between clean air groups and the gas industry grew even stronger between the two world wars. Capitalizing on this support, the industry increasingly used environmental claims to market gas and coke. Prefiguring much later campaigns to promote "green" products, one industry expert declared that an individual who heated with gas "is contributing to remove the nuisance of smoky skies which has for so long disfigured English towns. Gas is the sole practicable cure for this crying evil." As its leading trade publication explained in 1922, the gas industry, "both by its service and by its publicity campaigns," was doing much "to further the cause of Smoke Abatement and Coal Conservation." The industry received a major boost the following year when the chancellor of the exchequer, Neville Chamberlain, spoke at the opening of a national gas exhibition and praised gas for "relieving the lungs of the general public." Shortly thereafter, the executive chairman of the British Commercial Gas Association, completely overlooking its effects on gasworkers and the environment, told attendees at a smoke abatement conference that gas "involves no dirt or labour before, during, or after use."[56]

The National Smoke Abatement Society

On 5 October 1929, less than three weeks before the U.S. stock market crash, the leaders of the two largest antismoke groups in Britain made their way to the Peak District, a mountainous region between Manchester and Sheffield. Gathering at the Palace Hotel in the town of Buxton, they agreed to merge the thirty-year-old Coal Smoke Abatement Society (based in London) with the twenty-year-old Smoke Abatement League of Great Britain (based in Manchester) into a new group: the National Smoke Abatement Society.

Dr. Harold A. Des Voeux, who had long led the London group, chaired the meeting. Although the minutes report that he addressed those present on why he believed the groups should amalgamate, they do not even hint as to the reasons.[57] While the decision to merge may have been based entirely on a belief that a single large organization would have more influence than two smaller ones, financial difficulties may well have acted as a catalyst. The economy of Britain, as well as those of many other countries,

Figure 9.8. "Our Ideal: A Smokeless City." Note the gasworks on the lower right. Advertisement from Smoke Abatement League of Great Britain, *Report of the Smoke Abatement Exhibition Held at Bingley Hall, Birmingham, September 7th–10th, 1926* [1926], 68. Reproduced courtesy of Birmingham Library Services, Local Studies and History (391201).

experienced stagnation well before the dramatic events of Black Friday unfolded on Wall Street. Among the places first affected were the industrial cities of northern and northwestern England, southern Scotland, and Wales—places upon which the Smoke Abatement League of Great Britain depended for its membership subscriptions.

For most of its first decade, the amalgamated group was based in Manchester. This was a logical choice, since costs were considerably lower there than in London. According to a contemporary report, smoke emissions in Manchester had declined by 75 percent during the previous three decades, but the air remained very smoky. Over a thousand smokestacks stood in the city, in addition to the chimneys of 150,000 houses. Together, they poured forth smoke from three million tons of coal annually. As a result of this consumption, 200,000 gallons of coal tar were deposited on the city each year, along with 20,000 tons of ash and soot.[58]

During the economic crisis of the 1930s, the National Smoke Abatement Society received substantial support from the manufactured gas industry. The industry had a long history of using the issue of smoke abatement to promote itself, and many antismoke activists viewed the relationship as beneficial to both. In the early years of the Coal Smoke Abatement Society, Sir George Livesey had spoken at its conferences. With the creation of the National Smoke Abatement Society in 1929, the industry quickly sought to offer its support—and influence. In the early months of 1930 the Manchester District Institution of Gas Engineers urged gas companies to contribute funds to the society, and it immediately followed with a request to place a member of the gas industry on the executive board of the National Smoke Abatement Society. The society granted this request and issued a formal invitation to the gasworks manager the industry group had suggested.[59]

In 1936 the National Smoke Abatement Society organized a major exhibition on air pollution at the Science Museum in London. The exhibition, which attracted over 35,000 visitors, provided graphic evidence of the damage that smoke caused to vegetation, building materials, and the human body. Not for the squeamish, it even included specimens of human lungs, some of which were stained black with coal smoke.[60] In addition to demonstrating the harmful effects of coal smoke, the exhibition suggested that gas and coke provided the best solution to the problem. One part of the exhibition, produced by the British Gas Federation, noted that a million and a half gas cooking stoves and nearly a million gas heaters were in use within Greater London and asserted that they substantially reduced the amount of soot deposited in the metropolis. Another display highlighted the presence in coal of valuable chemicals, which were wasted when coal was burned in the home, but which could be recovered and used productively when coal was processed in gasworks. According to the official handbook to the exhibition, "The Gas Industry has a valuable contribution to make in the reduction of smoke and corrosive substances which enter the atmosphere." One year later, the Science Museum granted the British Gas Federation 4,000 square feet of space for a permanent exhibit to tell the story of manufactured gas from its origins to the present. The industry continued to drive home the message that its products were the solution to the smoke problem by helping to finance a documentary film entitled "The Smoke Menace" in 1937 and a booklet called "Britain's Burning Shame," which the NSAS published the following year. In the wake of all this activity, the *Gas Bulletin* boasted that "the gas industry has helped to put Smoke Abatement on the map; and in the process it has achieved for itself no little prestige."[61]

By the late 1930s, perhaps fearing that the gas industry was exploiting their group for its own business purposes, the NSAS began to view its offers of assistance with greater circumspection. In September 1938, just one year before the Second World War broke out in Europe, the Gas Light and Coke Company approached the society about cosponsoring an exhibit that the company had produced for the Charing Cross tube station in London. Although the NSAS agreed to the offer, the minutes emphasized that the exhibit was not a venal advertisement. Instead, it "was of a striking character . . . had cost a considerable sum to construct," and "been designed by Mr. Misha Black, a leading exhibitions designer." Most important of all, the minutes noted that the exhibition "dealt only with the effects of smoke and there was no reference to the use of gas in preference to any other smokeless medium." The society agreed to the proposal, and in 1939 thousands of people had the opportunity to see the exhibit. Following a tradition established by Victorian smoke abatement advocates, the exhibit stressed both the economic costs of smoke (wasted fuel, extra cleaning, transportation delays) and the health effects of lost sunlight. During the 1930s and 1940s the National Smoke Abatement Society also produced a series of posters that communicated these messages.[62]

Between its origins in 1813 and its nationalization in 1949, the gas industry in Britain provided many benefits, including light, heat, convenience, and fewer smoky chimneys from coal-fed stoves and fireplaces. During the first half of the twentieth century the average number of hours of sunlight rose in London and other cities, the frequency of dark smog decreased, and the quantity of soot deposited on surfaces diminished. The reasons for these improvements in air quality are complex. Although total coal consumption declined somewhat during this period (particularly during the Great Depression of the 1930s), coal continued to provide most of the energy consumed in Britain's factories and houses. Some of this energy came directly from the burning of raw coal; the rest came indirectly in the form of gas and electricity, produced in power stations and coal gasification plants that emitted less smoke per ton of coal than did household fireplaces. The change, coupled with a decline in urban density and improvements in the design of industrial boilers, led to a noticeable decline in particulate emissions.[63]

Despite the benefits that they provided, so-called smokeless fuels also caused considerable harm to both people and the environment. In contrast to coal smoke, which was relatively widely distributed, the disadvantages of gas and coke were more localized. In addition to damaging the health of workers who produced them and filling the air with smoke and acidic vapors, the manufacture of gas and coke polluted the water and soil with highly toxic contaminants including cyanide, heavy metals, and carcinogenic organic compounds. The carbonization of coal shifted many of the detrimental consequences of energy from places where it was used to those where it was produced, from consumers to workers, and created

Figure 9.9. Gasholder in Greenwich, London. Standing south of the Millennium Dome and across the Thames from Canary Wharf, this massive gasholder is a highly visible reminder of the work once carried out on this site by the South Metropolitan Gas Company. Photo by the author.

forms of environmental contamination that will persist far into the future. Yet the belief that gas and coke were clean forms of energy—that they were, in fact, the solution to what many people in Britain considered the most pressing environmental problem of the late nineteenth and first half of the twentieth century—blinded many to the detrimental consequences of coal carbonization and allowed industry to neglect changes that would have reduced the damage it caused to health and the environment.

The construction of the Millennium Dome, built in the late 1990s beside the Thames in the Docklands region of London, required one of the largest environmental clean-up projects ever attempted. This location, the focus of Britain's inauguration of the third millennium, was heavily contaminated by toxic residues from the massive coal gasification works that had formerly occupied the site. Commensurate with the rest of the Dome's budget, the cost of remediation was staggering, and it kept rising as work progressed. By the time the toxic dust had settled, the government had spent £185 million and had excavated 200,000 tons of contaminated soil.[64] Sixty-eight former gasworks sites—each of them a potential hazard—have been identified in the London area; in Britain as a whole, there are at least two thousand sites.[65] Throughout much of the industrialized world, gasworks and coke plants are today recognized as among the most hazardous and widespread sources of polluted soil and groundwater.

CHAPTER 10

Death Comes from the Air

Normally healthy people should not be worried. If you breathe through the nose, as you should, the nose acts as a filter to remove the particles of soot and dirt from the air.
—*A London physician, 1952*[1]

ON 1 SEPTEMBER 1939, just under twenty-one years after the end of what was sometimes simply called "the World War," warfare returned to Europe on a massive scale.[2] As had happened between 1914 and 1918, the pressures of wartime production led to heavy coal use and the tabling of smoke restrictions. In some places, smoke was produced deliberately in an attempt to conceal cities and factories from the view of enemy bombers. More significant, however, was the smoke that resulted from the wide-spread practice of maximizing industrial output by forcing steam boilers to exceed their design capacity. James Law, a combustion engineer from Sheffield, asserted in the middle of the war that in thirty years of professional experience he had "never seen conditions worse than those in the industrial cities and towns of the north during the past two years." According to him, instances of excessive smoke emissions had increased sevenfold since the start of the war.[3] In contrast to this increase in industrial emissions, household smoke—which in many places accounted for most of the smoke—probably declined during the war because domestic coal supplies were subject to strict rationing.

In 1940, with London under heavy aerial bombardment, the National Smoke Abatement Society evacuated its offices there for temporary quarters in Surrey. Despite this move, the group remained active throughout the following five years of war. Soon after its relocation, the group published a booklet entitled *Smoke Abatement in Wartime,* which sought to demolish the assumption that dense smoke might help to cloak cities from aerial attack. Although it admitted that "under certain atmospheric conditions the palls that form over the towns and drift across country would make it more difficult for hostile aircraft to recognise specific objectives," it maintained that smoke caused at least as much impairment to defending aircraft and ground-based artillery. From a distance, moreover, smoke acted not as a screen, but as a signal: "Palls of smoke, forming clouds or 'smudges' of haze on the horizon would frequently make it easier to hostile aircraft to detect the presence of distant towns and industrial works, while the long plumes of smoke frequently seen, even in country districts would quickly betray the presence of isolated factories and works that might otherwise have remained unnoticed." The booklet also argued that because smoke was a sign of wasteful combustion, "the encouragement of smoke abatement is of direct assistance to measures for coal conservation and fuel economy that are to-day more essential than ever."[4]

In April 1942 chief alkali inspector, W. A. Damon, predicted that "economic conditions will continue to be severe for many years after the cessation of the present hostilities and it is probably [*sic*] that there will be many problems of reconstruction more important than the conversion of furnaces to the burning of smokeless fuel. Nevertheless it would be well to determine post-war policy in this connection even though it might not be practicable to implement it for some time."[5]

As the end of the war became a foreseeable reality, talk of postwar reconstruction picked up steam. Determined to shape the direction that this took, in November 1943 the National Smoke Abatement Society held its first conference since the outbreak of war over four years earlier. The keynote address, by the writer and open spaces activist Clough Williams-Ellis (1883–1978), made repeated comparisons between the struggle for clean air and the fight against fascism. Speaking of smoke, but using language that his audience had often heard in relation to the military conflict, he maintained that the battle could not "be won by the single combat of heroes but only by a people's army, led by heroes certainly, but itself filled with zeal and a conviction that its cause is good." In a clear allusion to occupied Europe and the anticipated D-Day assault (which was to occur exactly seven months later), he noted that the National Smoke Abatement Society "has conducted a long-drawn guerilla warfare against the enemy, it has won various quite important engagements, but it has by no means won the *war*. That, I am convinced, will only be won by a major frontal assault with overwhelming forces at a favourable time. The most favourable time for winning an anti-smoke war will be almost immediately after the anti-fascist war, and it is for that zero hour that our maximum forces must be ready."[6]

To this end, the society issued numerous public reports and submitted policy papers to government departments that called for attention to the smoke issue. One such document, issued in 1942, declared that postwar reconstruction would require large numbers of "fuel-burning appliances of all kinds. With reasonable foresight and control modern technology can ensure that practically the whole of this shall be smokeless in operation. Without such foresight and control an entirely unnecessary volume of pollution will add to and help to perpetuate the existing problem."[7]

By the early 1950s the air over London was still far from pristine, but many assumed that the dense smog that had enshrouded the city in the late nineteenth century could never return. Far more people were concerned about obtaining adequate supplies of coal than worried about the smoke that it produced. The National Coal Board—a government-run monopoly established when Britain's entire coal industry was nationalized in 1947—was stung by frequent complaints about its inability to meet consumer demand. As the winter heating season began in 1952, the board energetically promoted "nutty slack" for use in household heating. In contrast to ordinary coal, which would remain subject to rationing until 1958, the board announced with much fanfare that nutty slack would become exempt from rationing beginning 1 December 1952. In the view of some critics, this "small coal" could more accurately be described as "large dust." Nutty slack was not only "exceptionally filthy" to handle, but also extremely smoky. When the National Smoke Abatement Society circulated a draft report that criticized this type of fuel, the National Coal Board responded aggressively. Meeting with the technical committee of the society, an official with the board threatened that publication of the report "might affect the relations between the N.C.B., and the Society, which so far had been most cordial." Despite this pressure, the society decided to proceed with the publication of their report on nutty slack.[8]

Four days after the end of rationing of nutty slack, on 5 December, disaster struck—raising the weekly death rate in London to a level that

Announcement by the

NATIONAL COAL BOARD

After the 1st December you can have
as much Nutty Slack coal as you like

off the ration

20/- a ton (1/- a cwt.) cheaper than ordinary house coal

You can now buy as much Nutty Slack as you like. It's off the ration. Nutty Slack is much smaller coal than you are used to, but it burns brightly and warmly if you mix it sensibly with your ordinary house coal.

What is Nutty Slack ?
Nutty Slack is just what it is called —slack with nuts in it ranging in size up to lumps about as big as two match boxes put together. Your coal merchant will show you Nutty Slack and tell you how to make the best use of it.

NOT AVAILABLE IN SOME AREAS
Nutty Slack will not be available in Scotland, South Wales, Cumberland, Durham and Northumberland, but in these areas other off-the-ration coal may be available.

It is just as expensive to mine and to carry on the railways as ordinary house coal. But it is offered at 20/- a ton cheaper than ordinary (Group 4) house coal because it is not as clean and is smaller.

For open or closed fires
It is good fuel to use in an open fire or an openable heating stove, mixed with your ordinary house coal. It is perfect for banking up your fire at night or when you are going out. It is particularly useful in a modern continuous burning grate.

Order as much as you like
You can buy it from your own or any other coal merchant and you can have as much as you like all through this winter. But don't wait till all your ordinary coal allowance is finished before you buy Nutty Slack. Order your Nutty Slack this week in time for Christmas. Use it with your ordinary house coal and make sure you get through the winter in comfort.

Nutty Slack

will help to keep the home fires burning

however cold and long the winter

A large-scale enquiry carried out this year for the Coal Utilisation Council shows that :—
1. 98 out of every 100 families in Great Britain have a coal fire of some sort in their main living room.
2. 84 out of every 100 families have no other way of heating that room.
3. 44 out of every 100 families say they need more coal than they are allowed.

Nutty Slack will help

Figure 10.1. In an attempt to stretch fuel supplies, the National Coal Board encouraged household consumers to use low-quality (and highly polluting) "nutty slack." This advertisement appeared less than two weeks before the onset of the 1952 smog disaster. Source: *Times* (London), 25 Nov. 1952, 5.

had not been seen since the worst days of the Blitz (the German bombardment of London during the Second World War). This time, however, the catastrophe came not from enemy bombers, but from the metropolis itself. An unusual combination of meteorological conditions—temperature inversion, low temperature, and lack of wind—interacted with coal smoke from railway locomotives, factories, and millions of chimneys to blanket the metropolis with a dense mixture of fog and smoke that reduced visibility to virtually zero. Half of the devices that measured airborne particulates in London became so full that they could no longer function, and sulphur dioxide reached the highest concentrations recorded since detailed monitoring of this compound had begun in 1932.

Newspapers devoted considerable attention to the fog as it was occurring, but initial reports said little about its effect on health. Although one article reported that a number of Londoners had sought medical attention for "fog cough," it suggested that there was no cause for alarm.[9] Yet thousands of Londoners were about to die from what became the deadliest environmental catastrophe in modern British history. By the time it was over, at least four thousand people lay dead.[10]

The events of 1952 raised a host of questions: How many people became sick or died as a result? Why was this particular fog so fatal? Was exposure to "normal" levels of pollution dangerous? The answers given to these and other questions were crucially influenced by an interaction of scientific theory, politics, and statistics, which together acted as lenses through which people interpreted London's deadliest encounter with fog.

Fog and Politics

Newspaper reports during the 1952 fog said virtually nothing about its possible health effects. Most articles focused instead on the visual aspect of fog. Some newspapers, celebrating the ability of dense fog to transform the appearance of London, published ghostlike photographs that followed a long tradition of depicting fog-enveloped landscapes as mysterious and sublime.

Despite some positive coverage that portrayed fog as an interesting visual display, most articles suggested that this fog was unwelcome. Yet even the most negative coverage of the fog tended to consider it an exclusively visual phenomenon. Reflecting this understanding, the first article to appear in the *Times* about the 1952 fog carried the headline "Fog Delays Air Services." In addition to grounding nearly all flights in and out of

London (Heathrow) Airport, poor visibility caused the suspension of bus and trolley routes, forced boat traffic on the Thames to a standstill, and prompted the Automobile Association to warn motorists against driving until conditions improved. The *Times* reported that "those who ventured on to the roads in the gloom of what should have been daylight made little progress, and many had to abandon their cars and walk." Echoing nineteenth-century reportage of fogs, newspapers in 1952 reported that criminals were deftly exploiting the fog to conceal a rash of "burglaries, attacks, and robberies." On the second day of the fog, all five of the Football League games scheduled to be played in London were cancelled due to "a complete black-out," and Wembley Stadium was forced to remain closed for the first time since it had opened in 1923. Organizers of a cross-country running match in south London went ahead with their race, but fog restricted visibility for spectators and competitors alike to only ten yards, and runners quickly became "lost in the gloom." Even the BBC was affected, for a number of people who were scheduled to go on the air found it impossible to reach its studios. The fog did not remain confined to the outdoors, but seeped into houses, offices, and concert halls. A music critic, reporting on a concert held on 5 December near Oxford Circus, complained of difficulty in seeing the performance through the fog in the auditorium. Noting that the vocalist had struggled to reach high notes, the critic remarked that "perhaps the weather was to blame." On 8 December the fog inside Sadler's Wells Theatre became so intense that a performance of *La Traviata* was halted after its first act. The manager, who issued refunds to everyone in the audience, noted that such a closure was unprecedented in the history of that theater.[11]

For many Londoners the fog of December 1952 caused little more than irritated throats and transportation difficulties. For thousands of others, however, it made breathing impossible. The Emergency Bed Service, an agency that coordinated hospital admissions, experienced an unprecedented demand for hospital care during the fog. During the week ending 13 December it facilitated the hospitalization of 2,019 individuals, 73 percent more than the existing weekly record and more than twice as many as had been admitted under its auspices during the corresponding week of 1951. During the last day of intense fog 492 individuals applied for emergency admission, 102 of whom were turned away because hospitals were already filled to capacity.[12]

Although medical personnel noticed a sharp rise in health problems, public recognition of the disastrous effects of the fog was slow to emerge.[13]

Delayed reaction to the fog can be attributed in part to the time needed to assemble and analyze mortality statistics, but it also resulted from the ruling Conservative Party's attempt to escape blame for the tragedy and to avoid being pressured into stricter controls on air pollution. On 16 December a member of Parliament from the Labour Party asked the minister of health to reveal "how many persons died of bronchial or other ailments in the greater London area as a result of the recent severe fog." The minister, Iain Macleod, initially sidestepped the question by answering that complete data were not yet available. One day later, the same MP called for the creation of a committee to "study the causes and cure of London fog." A representative of the Ministry of Works bluntly rebuffed this proposal and asserted that "a committee, the Atmospheric Pollution Research Committee, already exists under the Fuel Research Board and includes representatives of all the interested departments." On 18 December, in response to further questions, this time from fellow Conservatives, the minister of health issued a shocking statement: "The number of deaths from all causes in Greater London during the week ending 13th December was 4,703 compared with 1,852 in the corresponding week of 1951." The next day, newspaper headlines screamed, "Fog Week Deaths Rose by 2,800." The *British Medical Journal,* not known for sensationalism, referred to this rise in the death rate as nothing less than "spectacular."[14]

In the weeks and months that followed the fog, the government attempted to minimize the magnitude of the calamity, but public interest and concern remained high. Six weeks after the fog lifted, a member of the House of Commons asked the minister of health to describe how his department was represented on the Atmospheric Pollution Research Committee. In contrast to the government's assertion a month earlier that "all the interested departments" were involved, Macleod admitted that the Ministry of Health was not in fact represented on the committee.[15] Criticism of the government's handling of the fog question continued to rise, and members of the Labour Party sought to portray it as inattentive and callous. This impression was reinforced two days later when the minister of health publicly complained that "he seemed to get nothing except questions about fog and its effect on people's health. 'Really, you know,' he said, 'anyone would think fog had only started in London since I became Minister.'"[16]

Future prime minister Harold Macmillan, then minister of housing and local government, also faced harsh criticism in Parliament. In late January 1953 one Labour MP asked him, "Does the Minister not appreciate that last month, in Greater London alone, there were literally more people choked to death by air pollution than were killed on the roads in the whole country in 1952? Why is a public inquiry not being held, seeing that inquiries are held into air and rail disasters which do not affect so many people?" Joining in, another Labour MP asked, "Is the Minister aware that his complacency in dealing with this problem is creating a lot of dismay. . . ?" Throughout December and January the government deflected calls for an investigation into the fog by claiming that this would needlessly duplicate the work of the Atmospheric Pollution Research Committee. On 12 February, however, this tactic was exposed as a diversion. Asked what action that committee was taking in response to the December fog, a government spokesperson replied, incredibly, "None. The Committee is essentially an advisory body concerned with the collection of data on atmospheric pollution and with the researches into its measurement and prevention." In response to further questioning, he stated that the committee had met only two times in the previous year. Shortly after this embarrassing revelation, the minister of health announced that his department had begun an inquiry into the fog—the results of which were not made public until 1954.[17] Although this move suggested that the government had begun to take the pollution issue more seriously, it failed to satisfy many critics. Addressing the Commons five months after the fog episode, Norman Dodds declared,

> There is still an amazing amount of alarm on the part of the public about the heavy death roll [*sic*] and the widespread sickness following the December fogs. This alarm has been greatly increased by the amazing, at least outward, apathy of the Government. Most people who have deep feelings about this just cannot understand why there has not been a public inquiry after thousands of people were choked to death during the December fogs.

Predicting that a future fog could kill even more, Dodds warned that London might again be filled with the plague cry, "Bring out your dead!" Responding to Dodds's speech, the deputy minister of housing and local government announced that the government would create a committee "to undertake a comprehensive review of the causes and effects of air pollution," not just in London, but throughout Britain.[18] In July 1953, seven months after the fog cleared, the government finally launched this investigation. It was led by the industrialist and engineer Sir Hugh Beaver

(1890–1967), managing director of the brewing firm Arthur Guinness and Co. and creator of the *Guinness Book of World Records*.[19] Many both inside and outside of government initially viewed the new committee as little more than a way to divert political criticism from the ruling Conservative Party. Shortly after the committee was set up, Harold Macmillan remarked to the home secretary that smog was "one of those things, like the floods, by which the efficiency of the Government is judged. There is nothing very much that we can do, but we can look as if we were doing it."[20]

Counting the Dead

The ability to see the fog of 1952 as a disaster depended crucially on the collection and interpretation of statistical data. In contrast to the generally straightforward matter of determining the number of injuries and deaths from disasters such as fires, floods, and train wrecks, determining how many people died from the fog of 1952 was a complex and imprecise task. Detailed records existed about the number of deaths that occurred in and around London each day and week, but deciding which times and places were relevant was a matter of interpretation. Decisions had to be made about the geographical scope of the inquiry: some looked at the Administrative County of London, home to 3.4 million people, which was ruled by the London County Council. Others considered Greater London, population 8.4 million, which included parts of the counties of Middlesex, Surrey, Kent, Hertfordshire, and Essex. Another important issue was the chronological scope of the inquiry: should it encompass only the immediate period of the fog, or subsequent weeks as well? Only after these decisions had been made was it possible to determine how many people had died during or following upon the fog.

Such a procedure did not, of course, distinguish between people who died because of the fog and those who would have died anyway from other causes. Although it would have been possible to analyze the health records of everyone who died in an effort to determine whether fog had caused their deaths, such an undertaking would have been massive and time-consuming. In addition, the exercise would have produced doubtful results, for a determination of whether fog was the decisive cause of death in a person with a chronic illness would have been fraught with ambiguity. Instead of examining each case individually, experts compared the total number of deaths during the fog with an estimate of the number who would have died in the absence of fog. Although such a procedure may sound simple, there was no clear way to accomplish it. Some statisticians used the week preceding the fog as their reference point, while others used data from the corresponding week of the previous year. A third option, designed to control for the possibility that the previous week or year was atypical, compared the number of deaths that occurred during and after the 1952 fog with an average based on the same period over the preceding five years.[21]

Two months after the fog episode, the *British Medical Journal* reported that the overall death rate in London during the week of the fog had been nearly as high as that during the severe cholera epidemic of 1866.[22] During the period 7–13 December 1952, the death rate in the Administrative County of London jumped to 2.6 times what it had been during the previous week. In Greater London, the death rate also rose sharply, increasing by a factor of 2.3.[23] Two particularly hard-hit boroughs of London, East Ham and Stepney, saw a fourfold increase in their weekly death rates.[24]

Several weeks after his bombshell announcement implying that 2,851 people had died as a result of the fog, the minister of health announced a new estimate that placed the death toll at approximately six thousand. This number was derived by comparing the number of deaths in Greater London during a five-week period that encompassed the fog of 1952 with the same period of the previous year.[25] Shortly after Macleod raised the official death toll, W. P. D. Logan, the government's chief medical statistician, argued for a smaller figure. Although he noted that the 1952 fog "was a catastrophe of the first magnitude in which, for a few days, death-rates attained a level that has been exceeded only rarely during the past hundred years—for example, at the height of the cholera epidemic of 1854 and of the influenza epidemic of 1918–19," he made the assumption that the effects of the fog were confined to a two-week period and focused his attention on the Administrative County of London, which contained five million fewer residents than Greater London. Logan admitted that it was "not possible to be sure . . . that all the deaths brought about by the fog had been registered within the two weeks ended Dec. 13 and 20. *Assuming that they were,* the number of deaths so caused can be estimated to lie between 3717 (the excess in the weeks ended Dec. 13 and 20 over the week ended Dec. 6, 1952) and 4075 (the excess in the weeks ended Dec. 13 and 20, 1952, over the corresponding average for 1947–51)."[26] Inexplicably, he failed to consider whether any of the deaths registered during the week ending 6 December were fog-related, even though the fog began on the morning of 5 December and the data he used showed that deaths that week exceeded the five-year average for corresponding weeks by 257. Even more significant is Logan's assumption, inherent in his calculations, that no deaths after 20 December should be attributed to fog exposure. Yet as the government's top air pollution scientist, E. T. Wilkins, pointed out, death rates remained high well after the end point of Logan's study. Between mid-December and mid-February, approximately eight thousand more people died in London than during the same period one year earlier. Wilkins observed that these deaths coincided with persistently high concentrations of air pollution. In his words, this constituted "a second pollution shock to Londoners whose powers of resistance may have been reduced by the earlier smog incident."[27]

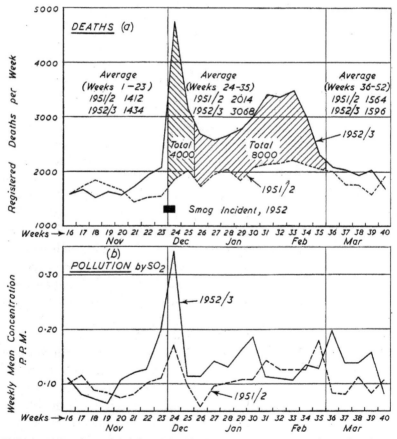

Figure 10.2. Deaths and Sulphur Dioxide Concentrations in London. This chart, prepared by E. T. Wilkins of the Fuel Research Station, was reproduced in Committee on Air Pollution, *Interim Report,* Cmd. 9011 (1953), 19.

Rejecting Wilkins's arguments, others maintained that influenza was solely responsible for the elevated mortality in early 1953, a view that was shared by the Ministry of Health. In light of the government's attempts to minimize the extent of the disaster, it comes as little surprise that it endorsed Logan's calculation of four thousand deaths as "the best estimate that can be made." A recent study, however, supports Wilkins's much higher estimate. Its authors, Michelle L. Bell and Devra Lee Davis, conclude that even if influenza deaths are excluded, approximately 7,700 more deaths than usual occurred in London during and soon after the fog disaster of 1952.[28]

Air Pollution and Health

In the aftermath of the 1952 fog, expert opinion was divided about the relative risk that polluted air posed to the general public. The minister of health, in effect blaming the victims, asserted that "the real problem of the 'smog' is that of persons—and particularly old persons—suffering from weaknesses of the lungs or heart."[29] In contrast, a report issued by the health committee of the London County Council emphasized that the increase in the death rate during the second week of December 1952, "although more pronounced among babies and the elderly, was not confined to persons of a particular age."[30] Echoing this perspective, the *British Medical Journal* noted that fog had nearly killed "a stalwart London policeman of 35 who collapsed and was admitted to hospital only just in time to save his life by means of oxygen and treatment for shock." In its view, "There is a danger that the obvious effects of the fog on babies and the elderly may divert attention from the damage possibly done to people in the middle years of life."[31] More stridently, the journal *Planning* insisted that "it was not, as has been occasionally suggested, merely a matter of old people, who were due to die soon in any case, being killed. Many of those affected might have had years of useful life ahead of them."[32]

Related to the idea that "normal" people, that is, healthy individuals in the prime of life, had emerged unscathed from the 1952 fog episode was the view that everyday levels of air pollution were acceptable. Attempting to reassure the public as the 1953 fog season approached, the minister of health declared that prolonged intense fog was "an extremely rare occurrence" and that it was "important not to confuse it with the more normal fogs with which we have long been familiar nor to exaggerate its effect on normal healthy people."[33] Criticizing this view, an earlier article in the *Times* had noted that although the 1952 fog "was admittedly an exceptional occurrence, it is obvious that even under less abnormal conditions the noxious matter in polluted air must have a detrimental effect on human and animal health and comfort; no doubt the effect is less violent, but all the more persistent."[34]

Statistical data lent credence to such concerns, as deaths from respiratory diseases associated with air pollution were many times higher in Britain than in much of the rest of Europe.[35] One of the most striking revelations to emerge in the wake of the fog disaster was the news that the death rate from bronchitis was many times higher in England and Wales than elsewhere in Europe.[36] Although this revelation shocked many members of the public, it was already widely known among health experts. According to statistics published by the United Nations, the bronchitis death rate in England and Wales was 62.1 per 100,000 people in 1952, compared to 18.3 in Belgium, 4.5 in France, and 3.4 in Denmark.[37] In addition to comparing Britain with other countries, the Beaver Committee noted that large cities in Britain experienced much higher death rates from bronchitis than did less polluted rural areas.[38] In his frequent public statements, Beaver also made historical comparisons in an effort to build support for clean air. He noted, for example, that respiratory diseases attributable to air pollution were killing a larger proportion of the British population in the 1950s than typhoid had killed in the 1870s.[39]

Initial reactions to the London fog of 1952 portrayed it as a natural disaster that was impossible to foresee or prevent. Fatalism and myopia may provide comfort in the face of tragedy, but they do so at the cost of concealing historical connections. Serious fogs *had* happened before in London, and their effects were in many ways similar from the mid-nineteenth through the mid-twentieth century. Throughout that period, the air of London was polluted with smoke, soot, and sulphur dioxide from millions of coal fires. Unusually still air and cool surface temperatures—the same conditions that made fog likely—caused pollution concentrations to rise to levels that interfered with breathing and proved deadly to thousands who suffered from bronchitis, asthma, and other diseases of the respiratory and circulatory systems. According to some experts, a combination of smoke and sulphur dioxide may be much more serious than either alone. Under normal conditions the nose and throat trap sulphur dioxide before it can reach more sensitive parts of the respiratory tract. When fine smoke particles are present, however, sulphur dioxide may adhere to them and be drawn deep into the lungs.[40]

Long-standing attitudes toward fog—particularly a tendency to view its effects as primarily visual—reemerged with the onset of fog in 1952. Yet public reaction to severe fog was markedly greater in the aftermath of 1952 than in the nineteenth century. Instead of viewing fog with a sense of complacency, many believed that it could be prevented and that government had a responsibility to see that it was. If deaths increased during foggy weather not because of cold air or high humidity, but because of the intense air pollution that often accompanied fog, this suggested that air pollution was unhealthy even in the absence of fog. Determining what had happened in December 1952 was a complicated and highly contested process, but one that ultimately led to the emergence of new ideas about air pollution and new policies to clear the air. What was initially seen as a natural disaster eventually came to be seen as a catastrophe that human beings had helped to create—and which they might also prevent from recurring.

CHAPTER 11

Smokeless Zones

The attainment of more complete combustion of the carbon does not . . .
reduce the quantity of the injurious invisible products poured into the air . . .
[such as] carbonic acid, sulphur, and chlorine compounds.
—*Harvey Littlejohn, 1897* [1]

Perhaps in another twenty years or so, it will be considered as great a social
sin to foul the air as to foul the water supply.
—*H. A. Des Voeux, 1936* [2]

ALTHOUGH MANY PEOPLE INITIALLY viewed the Beaver Committee
as engaged in little more than a disaster investigation, its mandate was
in fact much broader: to conduct a comprehensive study of "the nature,
causes and effects of air pollution," to examine the effectiveness of existing efforts to prevent it, and to suggest improvements. The committee's
efforts ultimately led to the passage of the Clean Air Act of 1956, which
tightened restrictions on industrial smoke and eventually required millions of people to stop using coal in houses and commercial buildings. [3]
The effort to ban most domestic coal fires required overcoming enormous
technological and cultural challenges. If people could no longer burn coal,
what were they to use for heating and cooking? Even if alternative sources
of energy were available in sufficient quantity (which was by no means
certain), most of them, including coke, gas, and electricity, required consumers to install new appliances. Were people willing to abandon lifelong
patterns of energy use for the common good? Could technology really
solve an environmental problem that had afflicted Britain's cities for well

over a century? This chapter addresses these questions by examining the conflicts that arose during the 1950s and 1960s as people in Britain debated whether, and how, to clear the air of coal smoke.

Following the 1952 disaster, many experts believed that the public needed to distinguish between natural and unnatural fogs. Seeking to educate its readers in this matter, the *Times* noted, "Fog is a natural phenomenon and is intractable by human agencies. Except for reducing visibility on sea and land, nature-made fogs are comparatively innocuous in themselves. It is when they get contaminated by man-produced impurities in cities and industrial areas that they become dangerous."[4] Sir Hugh Beaver similarly emphasized the distinction between natural and artificial fogs. Speaking at the University of London in 1955, he noted, "Although it is nature that produces the fog, it is we who produce the smoke. And this we have been doing for generations, to such an extent that we had almost come to accept it as natural and inescapable."[5]

Beaver's perspective echoed that of E. T. Wilkins. In a 1953 address to the Royal Sanitary Institute, Wilkins noted that "the bad effects of air pollution are, of course, not confined to periods of smog, for there is evidence that even the lower concentrations normal to many densely populated areas have persistent and insidious effects on public health, vegetation and . . . materials of all kinds. Thus the problem of smog is, in some respects, a short-term magnification of the general problem of atmospheric pollution." He added that "because the effects of normal pollution are ever present they undoubtedly represent, in the long run, a greater damage and loss to individuals and to the nation than does an occasional smog incident."[6]

After his committee had concluded its work, Beaver reflected that "we expressly avoided basing our arguments on the danger to health of particular incidents, such as the London smog of 1952. Not that we minimized that catastrophe in any way, but we felt that undue emphasis on it, would distract attention from the fact that damage to health and danger to life were going on all over the country, all the time, year in and year out." All of Britain, as he put it, constituted a "single permanently polluted area."[7]

Coming to grips with this problem was a daunting task, but one that the Beaver Committee accomplished with admirable efficiency. During its sixteen months of existence, the group convened 59 times as a full committee and held 74 subcommittee meetings. Though based in London, the entire committee met with local authorities, businesses, and nongovernmental organizations in many of the most polluted places in Britain, in- cluding Manchester, Glasgow, Leeds, Sheffield, and Birmingham. Just four months after its creation, the committee issued an interim report that summarized much of what was then known about the nature and effects of air pollution throughout Britain.[8] The committee argued that air pol- lution constituted a public health crisis that required a campaign as vigor- ous as the one that nineteenth-century sanitary reformers had waged for safe water. In addition to condemning air pollution for harming human health, the final report of the committee estimated that pollution caused £150 million in damage to textiles, metals, and buildings each year, and that it cost at least another £100 million in time lost to illness and trans- portation delays.[9]

Early in its deliberations, the committee agreed that coal combustion produced virtually all of the smoke in Britain's air. Contrary to popular opinion, it calculated that although houses and commercial buildings to- gether used less than 20 percent of the coal consumed each year in Britain, they produced 45 percent of the country's smoke. The reason for this was simple: fireplaces and household heating and cooking appliances gener- ally failed to burn coal as efficiently as did industrial furnaces. In its final report, the committee argued that since nearly half of the smoke in Britain came from houses and commercial buildings, these had to be in- cluded in any serious effort to improve air quality.[10] Although this may seem self-evident, it was a controversial statement at the time. Ever since the nineteenth century, national legislation had explicitly exempted do- mestic smoke from regulation.

In drafting its recommendations, the Beaver Committee was impressed by the success of recent efforts to establish so-called "smokeless zones," areas within which the creation of any smoke was prohibited. The Na- tional Smoke Abatement Society had suggested the idea in the mid-1930s, but war delayed its implementation until 1951, when Coventry established the first smokeless zone in Britain. Manchester followed suit the follow- ing year. Although its smokeless zone initially comprised only 104 acres in the center of the city, the effect on air quality was evident almost immedi- ately. During the winter of 1952–53, people in central Manchester experi- enced fewer days of smoke-filled fog and discovered that soot was accumu- lating more slowly than usual.[11] Building on this experience, the Beaver Committee recommended the formation of smoke control areas in the most highly polluted parts of Britain. Within such areas, private houses would not be allowed to produce any smoke whatsoever, and industry would have to refrain from emitting dark smoke. Because quantities of smokeless fuels were limited, the committee decided to focus first on what it called "black areas": industrial and residential places where large quantities of coal were consumed within a small area and which experienced frequent natural fogs.

Alternative Sources of Energy

The committee considered three major energy sources to replace coal for domestic heating: gas, electricity, and solid smokeless fuels. Although gas and electricity offered many advantages, Beaver noted that they were "much more expensive as space heaters in Great Britain than is solid fuel" and would likely remain so well into the future.[12] In addition, few people seemed willing in the mid-1950s to switch entirely to gas or electricity. As one member of the committee observed, "We found throughout the country a very general demand for at least one open solid-fuel fire."[13] In addition to providing heat more economically than gas or electricity, solid smokeless fuels could be used in existing fireplaces after only minor modifications. As the Beaver Committee was eager to propose reforms that would not be too expensive, the cost of appliance conversion was an important consideration. Converting a typical coal-burning fireplace to burn coke cost between £3 and £5, but converting it to burn gas cost between £10 and £20. Another reason the committee opted for coke instead of gas or electricity as the main source of energy to replace coal was that large quantities of it could be stored for periods of high demand during cold weather.[14]

The term *solid smokeless fuels* actually referred to three separate products: anthracite, manufactured fuels, and coke. The first is a naturally occurring type of coal that at the time made up about 10 percent of Britain's coal production. In contrast to bituminous coal, which contains a high concentration of smoky volatile matter, anthracite is almost pure carbon. Because it contains little volatile matter, anthracite produces virtually no smoke. Britain possessed far too little anthracite, however, for it to become the main source of fuel in houses and commercial buildings. Manufactured fuels, on the other hand, gave the look of an ordinary coal fire without producing smoke. Created by the carbonization, or heat-treatment, of powdered bituminous coal, they were formed into briquettes that were easy to ignite. Marketed under names such as Coalite, Cleanglow, Burnbrite, and Homefire, they could be used in fireplaces designed to burn ordinary coal. Yet the high cost of manufactured fuels proved a major hindrance to their adoption.

After weighing the alternatives, the Beaver Committee concluded that most of the coal they wished to see phased out should be replaced by coke, a smokeless solid fuel made from coal. Doing so involved solving a number of technical challenges. As Dr. G. E. Foxwell, an expert on fuel and member of the Beaver Committee, noted, switching from coal to coke would require millions of new grates and appliances. Furthermore, coke was unpopular, expensive, often poor in quality, and too scarce to replace the large quantity of bituminous coal used for household heating and cooking. Additional coke would need to be freed up for domestic consumption by converting coke boilers in government and industry to burn oil instead.[15]

In recommending coke, the Beaver Committee faced an uphill battle. During the spring of 1954 it met with local government leaders in many of Britain's largest cities. In Edinburgh, the committee was told that it would take fifty years to convert all of the city's old appliances to ones capable of burning coke. The conversion of appliances was far from being the biggest challenge, however. Edinburgh's officials complained that coke was more expensive and dirtier than coal. Officials in Salford expressed similar reservations about coke, noting that "there would have to be a good deal of propaganda to educate the public on the merits of smokeless fuel before compulsion could be contemplated. There was a strong objection to coke. Coke was coal with the goodness taken out, and it was difficult to get over this argument."[16]

The Beaver Committee proved curiously evasive when it came to estimating smoke from the three coal-consuming industries it counted on to supply the country with smokeless forms of energy: electric power plants, coke ovens, and gasworks. In contrast to its quantitative estimates of the annual production of domestic smoke (900,000 tons), factory smoke (800,000 tons), and railroad smoke (300,000 tons), it merely asserted that their processes produced "small" amounts of smoke. It may be that the committee, in the interest of presenting an unambiguous case against domestic use of coal, downplayed the smoke that resulted from producing alternative sources of energy.[17]

Pollution Control

In the 1950s it was estimated that over one-third of the gas produced in Britain came from works using antiquated processes that released large quantities of smoke into the air. Only 30 percent of the gas sold by the North Thames Gas Board was produced in relatively clean continuous-process vertical retorts.[18] Perhaps this is why, in a revealing address to his industry colleagues, A. E. Haffner, the chief engineer of the Southern Gas Board, asserted in 1959 that the gas industry had been "treated gently" by the Beaver Committee. In contrast to the committee's assumption that gas was synonymous with clean air, those who lived or worked near the gas industry's operations knew that the gasification of coal was highly polluting. Haffner noted that gasworks were increasingly becoming "objects of public criticism" and that complaints and legal action would likely intensify as towns and cities throughout Britain established smoke control areas and people became more conscious of sources of smoke within their communities. Although he asserted that the gas industry was working hard to minimize smoke, he cautioned that "there were occasions when there was some danger of the dark smoke limit being exceeded."[19]

Concerned about the potential of some local authorities taking a tough stance against smoke from gasworks, in 1957 the industry had petitioned the government to exempt them from local control and instead regulate them through the Alkali Inspectorate. The government agreed, and in 1958 gasworks, along with coke plants and electrical power stations, became subject to air pollution regulation under the Alkali Act. In the years that followed, the inspectorate threatened on several occasions to shut down manufactured fuel plants that released high levels of pollution into the air.[20]

Local authorities almost invariably objected to giving up their regulatory authority over industry to the Alkali Inspectorate. Some asserted that the agency was too friendly toward industry, while others argued that local inspectors would be able to keep much closer tabs on polluters than would an inspector employed by the central government who visited only occasionally.[21] At a 1955 meeting between officials from the national government and members of city councils, one alderman complained that the Alkali Inspectorate had interfered with Sheffield's efforts to crack down on industrial smoke pollution. The minutes note that there followed "a rather uncomfortable interchange between the Chief Alkali Inspector and a number of the Councillors," after which the minister of housing demanded that the aldermen "provide chapter and verse if they wanted to attack his Alkali Inspectorate."[22]

Although the Beaver Committee sought to examine air pollution in its entirety, it confined most of its reports to discussion of visible smoke. The committee acknowledged, however, that sulphur dioxide was another serious form of air pollution, one that had played a large role in causing the deaths in London during the 1952 smoke emergency. In contrast to smoke, which could in theory be prevented, the burning of coal, as well as coke and oil, inevitably produced sulphur dioxide. During the 1950s coal combustion alone propelled over 5 million tons of sulphur dioxide into the air of Britain each year. Members of the committee held sharply differing views about what ought to be done about sulphur dioxide. In a private letter, Beaver revealed that the group had debated this question "hotly and at length." Although the committee eventually reached a consensus, its recommendations in this area were extremely cautious: new power stations in densely populated areas should install scrubbers, but only if 90 percent of the sulphur dioxide could be removed.[23] Despite the limited nature of this recommendation, it was fiercely opposed by the British Electricity Authority, which claimed that it would force the price of electricity up by 12 percent. The committee admitted that the removal of sulphur would cost money, but it estimated that it would lead to a price increase of only 4 percent.[24]

Regardless of how expensive such measures would prove to be, it was clear that they would be difficult to implement. Only one coal-fired power station in Britain then washed its flue gases to remove sulphur dioxide. An internal government memo noted that "flue gas *washing* is practicable only at power stations sited on the coast or on large rivers." Another option, "flue gas *cleaning* . . . needs large quantities of ammoniacal liquor, which is a by-product of gas works; but enough liquor is produced to clean the gases of only six large power stations." In fact, the washing of flue gases had contradictory effects. Although it certainly improved air quality, this came at the price of polluting the water with large quantities of sulphurous acid.[25]

Interestingly, although the Beaver Committee acknowledged that electrical power plants released sulphur compounds into the air, it implied that gasworks and plants that produced manufactured fuels did not. "The increasing use of town gas, a fuel virtually sulphur-free, will assist in reducing sulphur pollution, and the use of electricity will have the same effect when the sulphur problem at power stations has been solved."[26] Yet according to industry estimates, gasworks released 4.6 pounds of sulphur into the air for every ton of coal that they consumed. Hundreds of gasworks each emitted thousands of pounds of sulphur into the air every day.[27]

The committee's final report, issued in November 1954, called for an 80 percent reduction in coal smoke over fifteen years. To accomplish this, ambitious legislation would be necessary, legislation that the committee suggested be called the "Clean Air Act."[28] The report attracted enormous attention, both from the popular press and from professional groups. As one would expect, reaction to it differed according to the interests (and self-interests) of particular groups. The *British Medical Journal* lauded the report, proclaiming that "air pollution is the almost unchecked environmental evil of the age. Our knowledge and practice of the requirements of health in respect of food, water, and shelter are vastly in advance of those in respect of the air we breathe."[29] The Institute of Fuel was more critical. At a meeting it convened to discuss the report, one participant argued that bituminous coal could be burned without smoke if consumers would only exercise care to add a little coal at a time to their fires. To drive the message home to consumers, he suggested employing the methods of subliminal advertising. At little expense, the BBC could be provided with a record containing the words "little and often" in endless repetition, which it could broadcast during intervals between its broadcast programs.[30] Another speaker remarked that in a world in which "everything that is nice and good is either illegal, immoral, or fattening, I refuse to be deprived of some of the things that are dear to my heart, and one of them is the open fire." He went on to say that sitting before the flickering flames of a coal fire was similar to "watching the waves breaking on the sea shore—and who wanted to sit and look at the electric radiator or the gas fire?" Speaking at the same meeting, the female mayor of Tottenham suggested that attitudes toward coal fires were shaped by gender. Women, claimed Mrs. A. F. Remington, viewed the coal fire from a more practical perspective than men. "In almost every home," she explained, "it was papa who wanted the comfort of the open fire when he returned from work, and it was mamma who wished that electricity and gas were cheaper so that she could avoid some of the dirt caused by the open coal fire."[31]

The Clean Air Act

Almost immediately after the Beaver Committee issued its report, the Conservative backbencher Gerald Nabarro (1913–73) announced that he planned to introduce a bill that embodied many of the committee's recommendations. His fellow MPs strongly supported his efforts and ranked it first among private members' bills that they hoped to see introduced in the coming session. Because this bill was not sponsored by the government, parliamentary rules prevented it from containing provisions that would require the expenditure of public funds. The National Smoke Abatement Society supported Nabarro's efforts and pledged £100 to support the administrative costs of the bill. The society made it clear, however, that Nabarro's bill was not enough. On 27 January 1955 its executive council unanimously approved a resolution that thanked Nabarro for the "stimulus" his bill was providing, but also urged the government to introduce a comprehensive clean-air bill.

At the same time that Nabarro was putting together his bill, the minister of housing and local government, the secretary of state for Scotland, and the minister of fuel and power urged the government to introduce legislation to implement the committee's recommendations. "We think," they argued in a memorandum to the Home Affairs Committee of the Cabinet, "that public opinion is ready for a strong Government lead and would support measures on the scale proposed. Indeed we do not think that anything less than positive action on a national basis would satisfy public opinion, or prove effective in abating pollution." They noted, however, that cleaning the air would involve surmounting significant social and technological obstacles, including "problems of fuel supply, a radical change in domestic heating habits, and large-scale expenditure on converting appliances in people's houses."[32] Spurred by these arguments, as well as external pressure from the Royal Sanitary Institute, the Federation of British Industries, and representatives of local government, the government introduced a bill of its own in 1955.

Although the government announced that it accepted the broad principles contained in the Beaver report, it adjusted many of the details. The Beaver Committee had recommended that the national government reimburse owners or occupiers in smoke control areas 50 percent of the cost of replacing old household coal-burning appliances with ones that could use smokeless fuel. The government's bill reduced the size of this subsidy, leading Beaver to protest in a confidential letter to Duncan Sandys, the minister of housing and local government. In its final form, the act required owners or occupiers of private houses to pay 30 percent of the cost of adapting fireplaces to burn smokeless fuel. Local authorities would pay another 30 percent of the cost, and the national government would contribute the final 40 percent.[33]

The Beaver Committee had recommended that clean air be declared national policy, and when the government's bill failed to do so, Beaver personally asked the minister of housing to add language to that effect.[34] Such a statement would have far more than symbolic significance, for Britain's nationalized industries—which included coal, gas, and electricity—were legally required to follow national policy. Had the National Coal Board and other government departments followed such a policy before the 1952 smog episode and not promoted the sale of nutty slack for household use, many lives might have been saved. Several years later, a similar conflict between clean air and other goals occurred when, to protect Britain's coal industry and hold down its growing trade deficit, central government policies restricted imports of coal and oil and pressured Britain's nationalized electrical utility to build coal-burning power plants rather than nuclear ones.[35]

Shortly after the government released its draft bill, a group representing the coal industry objected that it might encourage people outside of smoke control areas to needlessly stop using bituminous coal in their houses. To prevent this from happening, the Chamber of Coal Traders suggested exempting private dwellings from a clause in the bill that prohibited dark smoke.[36] Considering that both gas and coke came from coal, why was the coal industry upset about losing domestic customers? It was a question of economics: suppliers earned more profit per ton selling coal to domestic consumers than to factories. In London, during the summer months when coal was cheapest, domestic consumers still paid up to 75 percent more for a ton of coal than did gasworks along the Thames.[37]

The National Smoke Abatement Society worked hard to counter the arguments of the coal industry. In a pamphlet that explained the rationale for different requirements applying to industrial and nonindustrial smoke, for beginning the program on a small scale, and for the change in terminology from "smokeless zone" to "smoke control area," it noted,

> Even with the most modern and best-run [industrial] plant, a little light smoke is at times unavoidable, and a slight haze may in fact accompany the highest combustion efficiency. For all practical purposes such smoke can be tolerated, and is, in fact, allowed in the new Act. But it can hardly be allowed in a smokeless zone without making nonsense of the phrase. It was for this reason that the Beaver Committee proposed the term "smoke control areas."[38]

Despite widespread interest in clean air, some expressed reservations about allowing the government to police what people did within their own houses. One of the most outspoken critics of such proposals was the *Economist* magazine, which published an article in 1955 entitled "Gestapo of the Grates." Coming only a decade after the defeat of Nazi Germany, the suggestion that the British government might force its way into houses to see if people were engaged in prohibited activities touched a nerve. In July 1955 the minister of housing informed Beaver that the government had altered its draft bill to make it clear that the government would not search private dwellings for contraband fuel. This step was necessary, he warned, to avert "serious antagonism . . . on the lines of the recent article in *The Economist.*"[39] The National Smoke Abatement Society shared this concern, concluding that "it would not be advisable to press for right of entry to private dwellings to be given. The more effective way of dealing with domestic smoke emissions in smoke control areas would be by the prohibition of the sale of any but authorized fuel."[40]

Four years after the great smog killed thousands of Londoners, Parliament finally took the step that it had so long resisted: it began to restrict the use of coal in people's houses. By the time the Clean Air Act passed in 1956, twenty localities had already obtained special parliamentary authorization to establish smoke control areas, and ten had put them into practice.[41] Following the passage of the law, the national government ordered local authorities in over three hundred so-called "black areas" to submit plans for establishing similar areas within their jurisdictions. With the exception of London, almost all of these places were located in the industrial and coal-mining areas of central and northern England. One such place was Salford, an industrial city adjacent to Manchester. There, the local health department embraced the process. A pamphlet that it issued to residents began, "Dear Fellow Citizen, Your home . . . is in an area where the burning of smoke-producing fuel will shortly be made illegal." The document went on to explain the harm that smoke caused to health, buildings, and horticulture, and insisted that the benefits of clearing the air "outweigh by far the sacrifices needed to eliminate smoke."[42] Not everyone was quite so ready to embrace this sacrifice, however. As the National Smoke Abatement Society observed shortly after the act passed, "Although the majority of people today are likely to approve action for clean air in general, there may be resentment or resistance to the particular action that will affect them directly and perhaps call for a sudden break in the habits of a lifetime."[43] This prediction would prove all too accurate in the years that followed.

In a blunt report to Parliament published in 1960, the Ministry of Housing and Local Government complained that "the progress achieved by local authorities fell a good deal short of what had been planned" with

regard to the control of smoke.[44] Eighty-five of the 324 most polluted localities in Britain had not even developed a plan to deal with their smoke.
Twenty-one flatly denied that they had a problem, and thirty were reluctant
to prohibit the use of coal because some of their constituents—miners
and their families—received free coal as part of their wages or pensions.

Mining Communities and Air Pollution

Although environmental historians have devoted considerable attention
to the effects of the British mining industry on land use and water quality, they have largely overlooked the air pollution that occurred in coal
mining communities.[45] Mining is an energy intensive process, and large
quantities of fuel were required to heat ventilation furnaces and power
steam engines used to pump water and lift coal from deep underground.[46]

On the eve of World War I, Britain's collieries consumed 18 million
tons of coal, more than 6 percent of their entire output.[47] Not surprisingly, the air near coal mines was often filled with thick clouds of smoke
and sulphur dioxide—a problem that was compounded by the frequent
presence of smoldering spoil heaps. Mining also dispersed tons of grit and
dust from blasting, drilling, and cleaning operations, and promoted wasteful and highly polluting patterns of consumption in pit villages and towns.
As the *Times* put it in 1954, air pollution was extremely intense in the mining areas of south Yorkshire, "where all miners have their free coal allocation and where there is a tradition of burning a great deal of coal on the
domestic hearth."[48]

Leonora Murray, who grew up in northern England in the early
twentieth century, recalled that a "terrible fight against smoke and dirt
[was] necessary in any colliery district." Near the entrance to the "smoke-
belching pithead," miners and their families "lived often in two-roomed
homes planted in long black rows facing each other." Until pithead baths
became common in the 1950s, miners returned home at the end of each
shift covered with coal. As Murray noted, "The children bring in loads of
black mud on their boots from the rutted, unpaved roads, and the men
bring in coal dust. I saw a woman pouring away the water in which she
had just washed her husband; it was black as ink with a two-inch sedi-
ment of coal dust." Murray also noted that "miners' wives made pillow-
slips of black sateen because it was impossible to struggle with white cotton ones, which so soon became grey, however much they were washed."[49]
Although most people who lived near collieries viewed mining as dirty

and unhealthy—even to people who did not work in them—they also recognized that their own livelihoods depended either directly or indirectly on mining. Over generations, the identity of both communities and individuals became deeply interwoven with mining.[50]

The mining of coal, like that of other minerals, brought to the surface large quantities of materials that had little or no commercial value. In the case of coal mines, this included not only rocks, but also bits of coal that were too small to interest most consumers. James Law, the chief smoke inspector for the industrial city of Sheffield, noted in the 1930s that "a colliery boiler-house is to all intents and purposes an 'incinerator,' where unmarketable fuels are disposed of. The fuels are probably low in calorific value and high in ash and moisture content."[51]

Although the owners of many coal mines saw little use for poor-quality coals other than burning them, others dreamed of turning waste into money—particularly when coal prices were high.[52] Not unlike the meatpacking industry of late-nineteenth-century Chicago, in which Philip Armour and others found ways to turn a profit from nearly every conceivable part of an animal's carcass, colliers sought to do the same with everything extracted from their mines.[53] "Concessionary coal" contributed substantially to the smoke problem, for it accounted for about one-sixth of Britain's total domestic coal consumption; in mining districts, of course, its share was much greater. Sometime in the nineteenth century, mine owners began to convert unsold coal into cash by unloading large quantities of it on their workers as payments in kind.[54] By the early twentieth century, miners in Britain were receiving approximately five million tons of so-called concessionary coal every year. This "generosity" cost the owners of coal mines very little, but they did not extend it unconditionally. Miners who received free or discounted coal often earned lower piece rates than those in mines where concessionary coal was not distributed.[55] When Britain's coal industry was nationalized in 1947, it employed approximately 700,000 miners at nearly a thousand collieries.[56] All but about 50,000 of them were entitled to free or cheap coal.[57] Agreements about who was entitled to it and how much each person should receive varied from one part of Britain to another.[58] While miners in parts of Durham obtained as much as fifteen tons of concessionary coal per year in the 1940s, those in the rest of the county received an average of ten tons each.[59]

Concessionary coal varied not only in quantity but in quality. As the mining industry became increasingly mechanized during the 1940s and 1950s, "run-of-the-mine" coal became both dirtier and smaller. Between

1947 and 1962, the proportion of dirt in untreated coal tripled, and the proportion of small coal rose from 40 percent to 67 percent.[60] In 1956 a group of miners in Yorkshire complained that the concessionary coal they were being given was "rubbish." When they went on strike to demand that it be screened to eliminate dirt and small pieces, the National Coal Board responded by saying that it would do so only if the miners agreed to an individual allocation of only seven and a half tons each year. The miners remained steadfast, and after a strike of nearly two weeks, the Board agreed to supply miners with ten tons of high-quality coal.[61]

Both within mines and in the communities that surrounded them, little incentive existed to conserve coal. Per capita consumption of coal in nineteenth-century Britain was twice as high in districts where coal was mined compared to elsewhere in the country.[62] Although climate, culture, and cheap transportation no doubt played a role in this phenomenon, the availability of ample quantities of concessionary coal was arguably more important than anything else. Recipients of concessionary coal often received much more of it than they were able to burn in their own houses. Not surprisingly, many sought to exchange their extra coal for something that they could use—cash. In the eyes of the miners, pensioners, and widows who were entitled to concessionary coal, it seemed perfectly reasonable to sell it. After all, as employers noted whenever miners demanded higher wages, this coal was payment in kind. Yet mine owners regularly insisted that miners were not at liberty to do whatever they wished with this portion of their earnings, for the industry did not want its employees and their families to compete with it for customers. Miners who were caught selling their concessionary coal to others could lose their right to receive it and might even be fired.[63]

Concessionary coal proved to be one of the most complex and contentious issues facing those who wished to establish smokeless zones. In coal-rich Yorkshire, miners received around a ton of free coal each month. Concessionary coal, noted a resident of the county in 1957, "was regarded as part of a miner's income, and it was difficult to suggest that it should be surrendered altogether. The alternative was that it should be exchangeable for smokeless fuel—either solid fuel, electricity, or gas. The valuation placed on it by the miner was the market value for house coal of £7 or £8 per ton." Though the National Coal Board promised to buy back concessionary coal from any miner who lived in a place subject to restrictions on household smoke, it paid no more than £2 per ton, based on the rationale that concessionary coal was of a poorer quality than the coal that was supplied to paying customers. "The result of the discrepancy in valuation was that the miners' antagonism was carried into local government affairs, and considerable opposition was experienced in proposals for smokeless zones, or the introduction of tenancy agreements relating to the burning only of smokeless fuel."[64]

Many miners resisted the suggestion that they forego ordinary coal for some type of smokeless fuel. Rather than giving miners a quantity of smokeless solid fuel that was equal to the amount of coal that they already received, the National Coal Board offered miners either an annual payment of £42 or smokeless fuel worth this much. If they opted for the latter, miners would give up their existing allocation of concessionary coal in exchange for four and a half tons of manufactured solid fuel. Although the latter had a higher calorific value than coal, this amount provided the heating equivalent of only six tons of ordinary coal. While this might be enough for a single family, none would remain to sell to others.[65]

Although substantial reductions in household coal consumption occurred in the late 1950s and early 1960s, most were confined to London and other parts of southern England.[66] Greater London, which had been one of the smokiest places in Britain in the early fifties, reduced its household coal consumption by three-quarters between the midfifties and the midsixties. Levels of particulates in the air of London declined by almost the same amount during this decade.[67] In the north of England, the amount of household coal purchased from merchants declined by only 10 percent in the decade following passage of the Clean Air Act. An average household there burned three times as much coal as one in the southwestern part of the country. Not surprisingly, measurements of air quality revealed that the air in the north contained much higher concentrations of smoke than in areas where less coal was burned.[68]

During the 1960s, as smokeless zones were being established in many parts of Britain, some believed that the northern parts of the country had done so little to reduce air pollution because miners feared losing the profits they made by selling their concessionary coal. Sir John Charrington, who was both president of the National Society for Clean Air and a leading member of the smokeless fuels industry, charged that miners in Yorkshire sold more coal to the public than coal merchants did. Speaking at an air pollution conference in 1967, he declared that it was obvious that miners "are not going to vote for smoke control areas which will deprive them of their illicit profits." An official with the National Union of Mineworkers rejected this assertion as "a bloody slander." He admitted that the practice had once been widespread, but claimed that it had diminished significantly in recent years.[69]

Although Charrington's remarks implied that the government should crack down on miners who sold their coal under the table, it is arguable that if miners had been allowed to sell their concessionary coal openly to whomever they wished, this would have resulted in less pollution than under the existing prohibition. For although the market price of coal was about £8 per ton during this period, the National Coal Board usually gave miners only £1—and sometimes even less—for each ton of concessionary coal that it bought back from them. The prospect of selling one's extra coal on the black market for £4 a ton was certainly more attractive.[70] Miners who sold their coal illegally had a greater incentive to burn it efficiently, and thereby produce less smoke. Those who bought coal from miners, however, were likely to have the opposite reaction. Because they were purchasing it for half the regular price, they had considerably less reason to limit their consumption. If the NCB had paid miners more for their coal, the black market would have dried up, and both miners and the general public would have been more likely to use less coal. This was not in the interests of the NCB, however, for it sought to promote the use of its product, not discourage it.

Between the late 1950s and the late 1960s the amount of coal smoke in urban areas of the United Kingdom fell by 60 percent. Although both industrial and household smoke declined during this period, most of the improvement was in industry, which not only reduced its coal use, but also adopted new technologies that resulted in more complete combustion and installed pollution-control devices. As a result, the amount of smoke coming from industry fell roughly fivefold during this decade.[71]

Although houses and factories produced similar quantities of smoke in the early fifties, by the late 1960s about 85 percent of Britain's coal smoke came from houses.[72] In the United Kingdom as a whole, the total amount of coal used in houses fell by a third during the decade that followed the passage of the 1956 act. In absolute terms, the extent of this decline was almost equal for concessionary coal and that purchased through merchants. Relative to the number of miners employed, however, concessionary coal consumption actually rose during this period, from an average of 7.6 tons per year in 1956 to 8.9 tons in 1967.[73] It is important to bear in mind, however, that working miners were not the only people to receive concessionary coal. In many parts of Britain retired miners and widows of miners were entitled to it. From the 1950s through the 1970s, new labor agreements actually expanded the amount of coal offered to widows and pensioners,

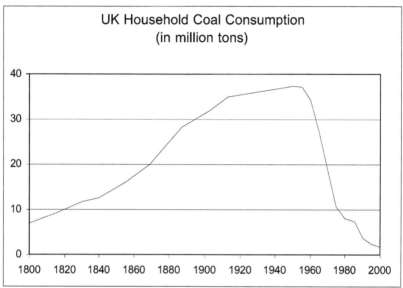

Figure 11.1. United Kingdom household coal consumption, 1800–2000. Sources: B. R. Mitchell, *British Historical Statistics* (1988), 259; idem, *Economic Development of the British Coal Industry, 1800–1914* (1984), 12; Department of Trade and Industry, *UK Energy in Brief* (2004), 17.

and extended the practice to areas in which it had not previously existed.[74]

Many recipients of concessionary fuel held tenaciously to their right to receive—and burn—real coal as opposed to a smokeless alternative. As a result, local officials in coal mining regions were extremely reluctant to enact smoke control regulations that would bar recipients of concessionary coal from burning the fuel that they received as part of their wages and pensions. Additionally, smoke control was viewed by many as a betrayal of a local tradition that was founded upon coal and an industry that was in decline. As late as 1975, three of every five miners in Yorkshire still used ordinary coal instead of smokeless fuel.[75]

Although coal-mining accidents rarely injured or killed anyone other than miners who worked underground, the health and environmental consequences of coal mining extended well beyond company property to affect not only colliery employees, but also their families, neighbors, and the environment. Tragically, what ultimately led to clean air in many mining communities was the collapse of the coal industry upon which they had been founded. As mine after mine closed during the 1970s and 1980s, most of the air pollution in these communities ceased, as laid-off workers lost not only their jobs and paychecks but also their free coal.

In 1956 British Coal, the name of the nationalized industry, extracted a quarter of a billion tons of coal and employed well over half a million mineworkers. In 1999 Britain's coal mines—which had returned to private ownership in 1994—produced 37 million tons of coal a year and employed only nine thousand miners.[76] Upon privatization, the government inherited the responsibility of providing miners who had made it to retirement age in the industry, and their widows, with the concessionary coal that had been promised decades earlier. About one hundred sixty thousand people continued to be eligible for concessionary fuel as the twentieth century came to a close. Many opted for cash in lieu of fuel, but the government continued to distribute about four hundred thousand tons of concessionary coal each year.[77] Although small amounts of coal are still mined in Britain, it is now cheaper in many cases to import foreign coal than to extract it from beneath the ground in the United Kingdom. After watching the British coal industry essentially disappear, retired miners found it bitterly ironic in the 1990s that the concessionary coal they were receiving was not mined in Britain.[78]

North Sea Gas

When Parliament passed the Clean Air Act in 1956, the vast fields of natural gas that lay beneath the North Sea had not yet been discovered. As a result, the government's clean air policy was based on the assumption that coal and its derivatives would provide almost all the nation's energy far into the future. Over time, fewer and fewer people were expected to burn raw coal in their houses, but the smokeless fuels they would use instead, whether gas, electricity, or solid smokeless fuels, were to come from coal.[79] In a speech on the country's energy future the previous year, the president of the Institute of Fuel had noted that natural gas exploration was occurring in Britain, but he concluded that little hope existed of a substantial find. Although he discussed several locations where exploration was occurring, he did not even mention offshore drilling as a possibility.[80]

In 1965 natural gas was discovered in parts of the North Sea claimed by Britain, and within a matter of years the country converted entirely from manufactured gas to gas from the North Sea.[81] Paradoxically, although the extraction of natural gas is much less polluting than the manufacture of gas from coal, this transition threatened to throw Britain's clean-air strategy into disarray, at least temporarily. As an internal government report had noted in the 1950s, the success of Britain's clean air efforts depended on a close "correspondence between the speed at which smokeless zones and smoke control areas are established and at which additional supplies of smokeless fuels become available."[82] If supply exceeded demand, the producers of smokeless fuel would resist expanding their production to accommodate future demand as the number of smokeless zones increased. The reverse could cause just as many problems. As a member of the Beaver Committee explained, "Nothing would be more disastrous than for a number of zones to be instituted as smokeless or smoke-controlled all over the country, and for them to prove failures because the right appliances had not been installed in advance and the right fuel was not available in sufficient quantities."[83] This fear proved well-founded.

The problem arose because gasworks not only produced gas, but were also the main source of house coke. As a result of decisions made in the 1950s, many houses in smoke control areas depended on coke as their primary source of heat. In recognition of this dependence, the national government provided incentives to encourage consumers of coke to switch to natural gas or electricity. In 1970 the sales director of the National Coal Board noted that recent shortages of coke had "caused . . . embarrassment to the government, local authorities, private industry and householders, and to the Board." For a time, the supply of coke was inadequate to meet demand, and the government was forced to postpone the implementation of new smoke control areas and temporarily suspend existing smoke control orders in others. Yet the low price of natural gas relative to coke soon made declining supplies of coke a moot point.[84] As natural gas grew in importance, the long-standing assumption that coal would continue to play a major role in supplying Britain with energy was abandoned. By the early 1970s some three thousand smoke control areas had been established in Britain.[85] Household coal consumption declined steadily and dramatically, falling to only one-sixth of its 1956 level by 1980. During that same period, the amount of gas sold in Britain expanded by a factor of six. Neither change would have been possible without the discovery of North Sea gas.

Attention to the difficulties and the success of the Clean Air Act of 1956 highlights not only the role of technological momentum in shaping environmental problems and responses to them, but also the challenges inherent in changing behavior one person at a time. The fact that Britain did clear its air of coal smoke proves that dramatic improvements can be achieved. By 1970, Britain's cities were largely free of the coal smoke that for so long had enveloped them.

Paradoxically, although the Clean Air Act of 1956 rid Britain's cities of their thick clouds of coal smoke, it fell far short of fulfilling the promise, contained in its title, of clean air. The long attention to the problems of coal combustion led many to overlook pollution from other sources, such as motor vehicles, which posed (and continue to pose) serious threats to both human health and the environment. In addition, since reformers had long conceived of air pollution in terms of visible smoke and soot, they tended to overlook the invisible gases that are released when coal—and other fossil fuels—are burned. The Clean Air Act of 1956 substantially reduced the amount of smoke entering the air, but it did little to reduce emissions of sulphur dioxide. Both the 1956 act and an amended act twelve years later placed heavy reliance not on reducing the amount of pollution that was created, but on more efficiently dispersing it into the environment. This strategy lessened the impact of certain pollutants in the places where they were produced, but it also shifted the burden to those downwind and failed to address the problem of pollutants that pose long-term global environmental problems.

Reinventing Pollution

FOR MUCH OF THE nineteenth century, Britain had the strongest industrial economy, most advanced technology, largest cities, and biggest factories of any country on the planet—all of which consumed enormous quantities of coal. Virtually every household in the country had a coal-fed kitchen range that was used year-round to cook food and heat water, and people who could afford them used additional coal fires to warm their homes during cold weather. Railroads and ships—even the London Underground—once were propelled by coal-fired steam engines. Large cities contained literally millions of domestic chimneys and thousands of industrial smokestacks. The decision to structure an entire economy and society around coal was a Faustian bargain, for although coal brought great wealth and power to some, made it possible for many to move at speeds faster than at any previous time in history, and provided warmth and comfort to virtually everyone, coal also filled the air with immense quantities of smoke and acidic vapors, which caused tremendous harm to human health and the environment.

As the first country to obtain most of its energy from fossil fuels, Britain was also the first place in which people sought to understand and alleviate the problems that these changes had unleashed. Although many people in Britain long considered coal smoke to be harmless—or even beneficial—during the late nineteenth century a radically different understanding of pollution emerged. Newspapers, magazines, and professional journals devoted great attention to coal smoke, and individuals in numerous communities came together to entreat householders, industrialists, and government officials to find ways to reduce or eliminate it.

The consequences of the epochal transition to fossil fuels continue to reverberate around the world. Global consumption of all three major fossil fuels—coal, oil, and natural gas—has expanded enormously in recent decades. In the developed world, this energy use now takes place without the emission of large quantities of visible smoke, but the situation is far different in places that are now making the transition from biological to fossil sources of energy. As the poor nations of the world develop, they are experiencing acute levels of coal smoke and other forms of localized pollution similar to those once found in Britain.

As I was researching and writing this book, people sometimes asked me to compare the extent of air pollution in the past with that of today. In trying to answer this question, I was acutely aware that the issue was more complex than it might at first appear, for pollution, as the anthropologist Mary Douglas has shown, is a malleable and historically contingent concept.[1] Although the idea of pollution has existed for thousands of years and is found in all human cultures, what counts as pollution in a particular time and place may not be looked upon as pollution in another. This is true not only when comparing vastly different cultural or historical contexts, but even when considering a particular society across relatively short intervals of time. Ideas about pollution, just like ideas about nature, are inventions of culture.[2]

Pollution and weeds have much in common. For those trying to eliminate them, it is self-evident that they constitute a nuisance. The mental step involved—from identifying a particular plant or substance to defining it as a problem—is often so automatic as to be left unrecognized. Both pollution and weeds, however, are socially constructed: they have no preexisting meaning. Perceptions of air pollution in Britain during the nineteenth and twentieth centuries were shaped not only by knowledge about the quantity of particular substances in the air, but also by the ways in which people understood nature, technology, and society. Skeptical of my attempt to situate pollution within a historical framework, one person interjected, "Forget all this postmodernist nonsense—just tell me data." "Fine," I replied, "but what type of data do you have in mind?" To talk about whether pollution has increased or decreased, one must decide what one means by it and where to look for it.

During the first three-quarters of the nineteenth century many people in Britain defined air pollution as miasma: gases produced by decomposing vegetables, garbage, corpses, and excrement. According to this view, natural processes rather than technological ones constituted the primary source of pollution. In contrast, smoke and other substances produced through chemical reactions were often seen as benign or even beneficial. Wild nature was suspected to be a source of deadly miasma, and coal smoke was frequently credited with providing an antidote to it.

Ideas about the relationship between human beings and the natural environment underwent a fundamental change in Britain during the late nineteenth century, a transformation that is exemplified by conceptions of and responses to air pollution. In the 1870s and 1880s, some began to conceive of technology rather than nature as the primary source of impure air. Concerns about the health effects of contaminants in the environment did not disappear but instead took new forms as people began to criticize the effects of technology on the environment, the body, and society. The natural environment went from being thought of as threatening and more powerful than humanity to something that was both pure and fragile. Many social reformers believed that nature would "civilize" the urban working class and reduce class conflict. The prevailing attitude toward technology, on the other hand, went from one of fascination and wonder to anxiety about the changes that it was causing to the environment and to society. Coal smoke, one of the most visible products of urban and industrial life, came to be viewed by many not as a welcome symbol of productivity, but as pollution. Indicative of the new attitude were the efforts of many people, both individually and collectively, to reduce or eliminate it. Adopting language and metaphors that had previously been confined to discussions of biological sources of disease, they strove to eradicate what some referred to as the "smoke plague."

Environmental reformers in the late nineteenth century, like many today, were divided between those who called for radical changes to the prevailing social and economic system and a larger group who advocated what may be called technological fixes. The former included individuals such as John Ruskin and William Morris who were convinced that coal smoke and modernity were inseparable—and who condemned both with equal ferocity. The latter included urban planners, who recommended wider and straighter streets to increase the movement of air in cities; sanitarians, who envisioned a system of underground pipes that would carry smoke away from its source and disperse it into the wider environment; and engineers, who offered technologies that purported to burn coal without producing smoke.

Although Victorian smoke abatement activists resemble later environmentalists in many respects, significant differences exist between them. Then as now, social critiques and environmental critiques were often connected: to fault the way in which a particular society interacts with the environment is either a direct or indirect criticism of that society. The particular social context and worldview of late nineteenth-century reformers fundamentally shaped the ways in which they understood the environment and human beings' relationship with it. These activists were motivated not only by concerns about the detrimental effects of smoke on health and the natural and built environments, but also by a belief that smoke intensified class conflict and promoted social disintegration, economic decline, and a loss of connection with the past and the natural world. Intertwined with these concerns was the perception that all of Britain was becoming subordinated to the demands of cities and that rural areas were losing their "natural" character. Smoke contributed to this by traveling long distances to defile areas that many wished to view as unaffected by cities, technology, and modernity. To many people it seemed that nature was in a state of decline: trees were becoming encrusted with soot and stunted by acid rain, dark smoke was reducing the amount of sunshine that urban areas received, and—half a century after Cruikshank created "London Going out of Town"—the countryside seemed more at risk than ever. Smoke eroded the distance not only between the city and the country, but between classes. In contrast to the dietary disparity between rich and poor, the air of a place affected everyone equally, regardless of social status. The only way in which the privileged could buy better air was by moving upwind from centers of population and industry—something that many of them indeed did.

The solutions that antismoke activists proposed differed depending on the ways in which they conceived of pollution. Those who disliked coal smoke because of an antipathy to modernity or industrialism were apt to focus their attention on smoke that came from factory smokestacks rather than on that from household chimneys. Despite the prominence of industrial smoke, however, a large proportion of the smoke that filled the air of Victorian Britain came from the cozy domestic fireplace. As a result

Figure 12.1. "The choice is yours," declared the illustration used to promote the 1926 Universal Smoke Abatement Exhibition in Birmingham. As this image suggests, many people then assumed that visible smoke was the only harmful product of coal combustion. From Smoke Abatement League of Great Britain, *The Universal Smoke Abatement Exhibition, Bingley Hall, Birmingham, September 6th–18th, 1926: Official Guide and Catalogue* [1926], cover. Reproduced courtesy of Birmingham Library Services, Local Studies and History (391200).

of their ideological predisposition against industry, many activists failed to recognize that smoke would not be banished by either rejecting or reforming the industrial system, but would require people like themselves to modify the ways that they used coal. Industrialists and their supporters, on the other hand, often agreed that smoke was a problem while denying responsibility for causing it. They complained that factories, in contrast to households, already faced legal restrictions on smoke emissions, and they argued that further mandatory reductions in industrial smoke were unfair and impossible to achieve. When confronted directly about the smoke that their factories produced, owners often blamed it on the alleged carelessness of the workers they employed.

Besides having different motivations, nineteenth- and early twentieth-century smoke abatement reformers defined air pollution very differently from later environmentalists. The former often considered coal combustion to be harmful primarily because of the unoxidized black particles that it produced. According to this perspective, the problem was largely one of smoke-shrouded skies and objects covered with soot and grime. Efficient combustion, they implied, would produce no pollution.

Now, more than half a century after the London smog disaster of 1952, the dark, choking clouds of coal smoke that obscured the skies of Britain's cities for generations are long gone. Most people in Britain heat and cook with natural gas, mainline railroads run on electricity, and branch trains and ships use diesel engines. Coal-fired boilers have disappeared from most factories, as indeed have many of the firms those factories once housed. Britain's coal mines, which long supplied coal to railroads and factories halfway around the world, have largely closed, the victim of the same capitalist logic that once created them. In the 1970s, for the first time in its history, Britain began to import significant quantities of coal.[3] Although coal mining was once the nation's largest single source of employment, providing work for a million men on the eve of World War I, only ten thousand people in Britain worked as coal miners at the end of the twentieth century.[4] If coal has lost the close connection that it once had to labor, so too has it disappeared from most people's perceptions of energy use. But although few people are aware of it, approximately 60 million tons of coal are still burned in Britain each year, and approximately a third of Britain's electricity still comes from coal.[5]

Despite Britain's success in ridding its air of pollution in the form of visible coal smoke, this achievement was in many ways a Pyrrhic victory. Consumption of fossil fuels in Britain, as in much of the world, has ex-

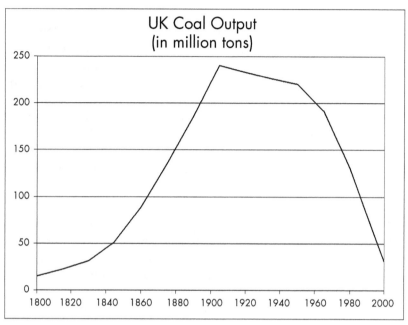

Figure 12.2. United Kingdom coal output, 1800–2000. Sources: B. R. Mitchell, *International Historical Statistics: Europe, 1750-1993*, 4th ed. (1998), 426ff; B. R. Mitchell, *British Historical Statistics* (1988), 247; Department of Trade and Industry, *UK Energy in Brief* (2004), 16.

panded enormously in recent decades. As a consequence, the products of fossil fuel combustion continue to pour into the air. Clean air reformers had long asserted that under ideal conditions, the combustion of high-grade coal would produce nothing more than two innocuous substances, water vapor and carbon dioxide. Although coal is known as a hydrocarbon, it consists of more than just carbon and hydrogen molecules. Coal also contains relatively large amounts of sulphur, as well as trace quantities of other elements, such as arsenic and mercury. When coal is burned, these substances are not destroyed, but either remain in the ash or go up the chimney.

Long after British smokestacks ceased to pour clouds of dark smoke into the air, they continued to emit massive amounts of sulphur dioxide. This gas, created when the sulphur in coal combines with oxygen during combustion, is a powerful acid that irritates the respiratory tract, harms plants and trees, and damages stone and iron. To prevent this from happening, engineers designed ever-taller smokestacks, which they hoped would propel sulphur dioxide high into the atmosphere and dilute it to insignificant concentrations. Tall smokestacks certainly made sulphur dioxide go away, but they did not eliminate it. Instead, they simply transferred British sulphur dioxide downwind, where much of it fell on Germany and Scandinavia. This unwelcome export, and the acid rain that it spawned, emerged as a contentious dispute between Britain and its European neighbors during the 1980s. Similar controversies erupted in North America, where large quantities of sulphur dioxide from the United States were being sent on the winds to Canada.[6]

As the amount of smoke and sulphur from coal declined in Britain's cities, the air became filled with pollution from another source, the internal combustion engine. Cars, trucks, and buses, their numbers growing every year as Britain became more affluent, saturated urban centers with carbon monoxide, sulphur dioxide, nitrogen oxides, lead, particulate matter, and volatile hydrocarbons. In yet another paradox, the increase in sunshine brought about by reductions in coal smoke interacted with vehicle emissions to form a new type of air pollution: photochemical smog.

Although first discovered in the Los Angeles basin, photochemical smog is now known to exist in cities around the world. One of its principal and most damaging ingredients, ironically, is ozone, the same gas that Victorian environmentalists hailed as the most important antidote to both miasma and smoke. Ozone exists naturally in the upper atmosphere, where it provides an essential shield against the sun's damaging ultraviolet rays—which were themselves once viewed as wholly beneficial, but have in recent years become linked to skin cancer. Ground-level ozone is a powerful irritant, and it can cause severe respiratory problems for individuals with asthma and bronchitis.

Many scientists now believe that the most serious consequence of burning fossil fuels is not smoke, soot, or even acid rain, but carbon dioxide. Produced when anything containing carbon is burned, carbon dioxide traps heat in the atmosphere. Concentrations of this compound have risen throughout the world since the onset of the industrial revolution—and continue to rise. Since no feasible way of capturing carbon dioxide as it is produced has been discovered, modern automobiles and coal-burning power plants release just as much carbon from a given quantity of fuel as do those that were built decades ago.

Nineteenth-century predictions that fossil fuel use could alter the composition of the earth's climate have become a reality. By 1995 approximately five thousand million tons of coal were being burned throughout the world each year, a five-hundred-fold increase since 1800. As a result of coal and oil combustion during the past two hundred years, the amount of carbon dioxide in Earth's atmosphere has risen by roughly 30 percent.[7] Humanity's impact on the global environment is more intense today than at any previous time in history. Unlike soot, particulates, and even sulphur dioxide, the effects of which are local and regional, carbon dioxide and the other greenhouse gases have the potential to alter the climate of the entire planet. If this happens, as a growing number of scientific experts warn is probable, the consequences for the world's environment and all who live in it may be devastating. Although the consequences of higher levels of greenhouse gases cannot be predicted with certainty, many scientists now believe that human-induced climate change has already begun.[8]

Ideas about what constitutes air pollution have changed enormously since the days of miasma theory, and they continue to be contested and reinvented. Today, as our understandings of air pollution are again changing, it is vital to know how they developed in the past. Pollution is a social construct, but the stuff that this concept signifies is very real indeed. To deal with it, we must understand not only its chemistry and effects but also the human attitudes, ideologies, and perceptions that brought it into existence.

Notes

Chapter 1

1. John W. Graham, *The Destruction of Daylight: A Study in the Smoke Problem* (London: George Allen, 1907), 143–44.

2. Arnold Marsh, *Smoke: The Problem of Coal and the Atmosphere* (London: Faber and Faber, [1947]), 264.

3. This phrase was coined by Peter Mathias, *The First Industrial Nation: An Economic History of Britain, 1700–1914* (New York: Scribner, 1969).

4. W. H. Te Brake, "Air Pollution and Fuel Crises in Pre-Industrial London, 1250–1650," *Technology and Culture* 16, no. 3 (1975): 337–59; B. R. Mitchell, *British Historical Statistics* (Cambridge: Cambridge University Press, 1988), 244.

5. Richard L. Hills, *Power from Steam: A History of the Stationary Steam Engine* (New York: Cambridge University Press, 1989).

6. Michael W. Flinn, with the assistance of David Stoker, *The History of the British Coal Industry,* vol. 2, *1700–1830: The Industrial Revolution* (Oxford: Clarendon Press, 1984), 252; Barry Supple, *The History of the British Coal Industry,* vol. 4, *1913–1946: The Political Economy of Decline* (Oxford: Clarendon Press, 1987), 15.

7. The quantities for all countries are expressed in metric tons. One metric ton equals 1,000 kilograms or 2,205 pounds. B. R. Mitchell, *International Historical Statistics: Europe, 1750–1993,* 4th ed. (London: Macmillan Reference, 1998), 426–35; idem, *International Historical Statistics: The Americas, 1750–1993,* 4th ed. (London: Macmillan Reference, 1998), 311–12.

8. B. R. Mitchell, *Economic Development of the British Coal Industry, 1800–1914* (Cambridge: Cambridge University Press, 1984), 1, 103.

9. William Stanley Jevons, *The Coal Question: An Inquiry Concerning the Progress of the Nation, and the Probable Exhaustion of Our Coal-Mines* (London: Macmillan, 1865); Graham, *Destruction of Daylight,* 139.

10. Mark Z. Jacobsen, *Atmospheric Pollution: History, Science, and Regulation* (New York: Cambridge University Press, 2002).

11. Between 1801 and 1901 the census registered a population increase in Glasgow from 77,000 to 762,000, in Leeds from 53,000 to 429,000, and in Sheffield from 46,000 to 381,000. Population growth appears even greater if one includes adjoining areas that were later incorporated into these cities. See Mitchell, *British Historical Statistics*, 26–29. For an excellent overview, see Asa Briggs, *Victorian Cities* (1963; repr., Berkeley: University of California Press, 1993).

12. London surpassed Beijing as the world's most populous urban area in the second quarter of the nineteenth century, and it retained this position until New York surpassed it in the first quarter of the twentieth century. See Tertius Chandler and Gerald Fox, *3000 Years of Urban Growth* (New York: Academic Press, 1974), 323–35.

13. James Johnson, *Change of Air, or the Pursuit of Health. . . .* (London: S. Highley, 1831), 1.

14. Mitchell, *British Historical Statistics*, 244–45; Roy Church, with the assistance of Alan Hall and John Kanefsky, *The History of the British Coal Industry*, vol. 3, *1830–1913: Victorian Pre-Eminence* (Oxford: Clarendon Press, 1986), 20; Stephen Mosley, *The Chimney of the World: A History of Smoke Pollution in Victorian and Edwardian Manchester* (Cambridge: White Horse Press, 2001), 16.

15. Te Brake, "Air Pollution," 339–41.

16. John Evelyn, *Fumifugium, or the Inconvenience of the Aer and Smoake of London Dissipated. . . .* (1661), repr. in *The Smoake of London: Two Prophecies*, ed. James P. Lodge (Elmsford, N.Y.: Maxwell Reprint, 1969). For an interesting recent interpretation of Evelyn's work, see Mark Jenner, "The Politics of London Air: John Evelyn's *Fumifugium* and the Restoration," *Historical Journal* 38, no. 3 (1995): 535–51.

17. Throughout much of the nineteenth century, the term *pollution* referred not to toxic substances, but to moral corruption and sin. *Oxford English Dictionary*, 2nd ed., s.v. "pollute," "polluter," and "pollution." Over time, however, the "environmental" definition of the word has banished the earlier definition to obscurity. For a discussion of this linguistic evolution, see Adam W. Rome, "Coming to Terms with Pollution: The Language of Environmental Reform, 1865–1915," *Environmental History* 1 (July 1996): 6–28.

18. Anthony S. Wohl, *Endangered Lives: Public Health in Victorian Britain* (Cambridge: Harvard University Press, 1983); I. G. Simmons, *An Environmental History of Great Britain: From 10,000 Years Ago to the Present* (Edinburgh: Edinburgh University Press, 2001).

19. Eric Ashby and Mary Anderson, *The Politics of Clean Air* (Oxford: Clarendon Press, 1981), 23. See also John M. Eyler, "The Conversion of Angus Smith: The Changing Role of Chemistry and Biology in Sanitary Science, 1850–1880," *Bulletin of the History of Medicine* 54 (1980): 216–34.

20. See J. E. Chamberlin, "An Anatomy of Cultural Melancholy," *Journal of the History of Ideas* 42, no. 4 (1981): 691–705, esp. 694–95.

21. See Martin A. Danahay, "Matter Out of Place: The Politics of Pollution in Ruskin and Turner," *Clio* 21, no. 1 (1991): 61–77.

22. J. Edward Chamberlin and Sander L. Gilman, eds., *Degeneration: The Dark Side of Progress* (New York: Columbia University Press, 1985); Pick, Daniel, *Faces of Degeneration: A European Disorder, c. 1848–c. 1918* (Cambridge: Cambridge University Press, 1989); F. H. A. Aalen, "Lord Meath, City Improvement, and Social Imperialism," *Planning Perspectives* 4, no. 2 (1989): 127–52.

23. Gareth Stedman Jones, *Outcast London: A Study in the Relationship between Classes in Victorian Society* (Oxford: Clarendon Press, 1971; repr., New York: Pantheon, 1984); Lynda Nead, *Victorian Babylon: People, Streets and Images in Nineteenth-Century London* (New Haven: Yale University Press, 2000).

24. Ashby and Anderson, *Politics of Clean Air,* 142; Z Archive, Z111/8, neg. 8605, Science Museum Library, Archives Collection, London.

25. Notable exceptions include B. W. Clapp, *An Environmental History of Britain since the Industrial Revolution* (London: Longman, 1994); James Winter, *Secure from Rash Assault: Sustaining the Victorian Environment* (Berkeley: University of California Press, 1999); and Ramachandra Guha, *Environmentalism: A Global History* (New York: Longman, 2000).

26. Raymond Williams, *The Country and the City* (Oxford: Oxford University Press, 1973); Graeme Davison, "The City as a Natural System: Theories of Urban Society in Early Nineteenth-Century Britain," in *The Pursuit of Urban History,* ed. Derek Fraser and Anthony Sutcliffe, 349–70 (London: Edward Arnold, 1983); Martin V. Melosi, "The Place of the City in Environmental History," *Environmental History Review* 17 (Spring 1993): 1–23.

Historians of the United States have produced much of the scholarship that integrates approaches from urban and environmental history. See William Cronon, *Nature's Metropolis: Chicago and the Great West* (New York: W. W. Norton, 1991); Christine Meisner Rosen and Joel A. Tarr, "The Importance of an Urban Perspective in Environmental History," *Journal of Urban History* 20 (May 1994): 299–310; Jeffrey K. Stine and Joel A. Tarr, "At the Intersection of Histories: Technology and the Environment," *Technology and Culture* 39 (Oct. 1998): 601–40.

27. See, for example, Joel A. Tarr, *The Search for the Ultimate Sink: Urban Pollution in Historical Perspective* (Akron, Ohio: University of Akron Press, 1996); Martin V. Melosi, *The Sanitary City: Urban Infrastructure in America from Colonial Times to the Present* (Baltimore: Johns Hopkins University Press, 2000); David Stradling, *Smokestacks and Progressives: Environmentalists, Engineers and Air Quality in America, 1881–1951* (Baltimore: Johns Hopkins University Press, 1999).

28. Two books on water pollution serve as admirable exceptions to the general dearth of scholarship on the interactions between social and environmental questions in nineteenth-century Britain. See Christopher Hamlin, *A Science of Impurity: Water Analysis in Nineteenth-Century Britain* (Berkeley: University of California Press, 1990); Bill Luckin, *Pollution and Control: A Social History of the Thames in the Nineteenth Century* (Bristol: Adam Hilger, 1986).

29. Raymond Williams, *Problems in Materialism and Culture* (London: Verso, 1980), 71.

Chapter 2

1. "Fresh Air in the Country," *Builder*, 23 July 1859, 488.

2. William H. Brock, *Justus von Liebig: The Chemical Gatekeeper* (Cambridge: Cambridge University Press, 1997), 203–4.

3. On historical conceptualizations of miasma, see Christopher Hamlin, "Providence and Putrefaction: Victorian Sanitarians and the Natural Theology of Health and Disease," *Victorian Studies* 28 (Spring 1985): 381–411; Margaret Pelling, *Cholera, Fever, and English Medicine, 1825–1865* (Oxford: Oxford University Press, 1978); Caroline Hannaway, "Environment and Miasmata," in *Companion Encyclopedia of the History of Medicine*, ed. W. F. Bynum and Roy Porter, 1:292–308 (London: Routledge, 1993); Sylvia N. Tesh, "Miasma and 'Social Factors' in Disease Causality: Lessons from the Nineteenth Century," *Journal of Health Politics, Policy, and Law* 20 (Winter 1995): 1001–24; Michael Worboys, "From Miasmas to Germs: Malaria, 1850–1879," *Parassitologia* 36, nos. 1–2 (1994): 61–68.

4. Quoted in John M. Eyler, *Victorian Social Medicine: The Ideas and Methods of William Farr* (Baltimore: Johns Hopkins University Press, 1979), 99.

5. "Chemical and Physical Modifications of the Atmosphere Consequent on Habitation," *Chemical News* 5, 15 Mar. 1862, 146.

6. Daniel Headrick, *The Tools of Empire: Technology and European Imperialism in the Nineteenth Century* (New York: Oxford University Press, 1981), 59–60.

7. J. Lane Notter, "Sanitary Notes," *Sanitary Record,* 15 June 1880, 445–49, esp. 447–48. For background, see Mark Harrison, "Tropical Medicine in Nineteenth-Century India," *British Journal for the History of Science* 25, no. 3 (1992): 299–318.

8. For an insightful discussion of nineteenth-century concerns about the exuberance of nature, particularly in tropical regions, see Donald Worster, *Nature's Economy: A History of Ecological Ideas,* 2nd ed. (Cambridge: Cambridge University Press, 1994), esp. 44–48. On ideological connections between imperialism abroad and elites' attempts to control the lives of the lower classes within Britain, see Dorothy Porter, "'Enemies of the Race': Biologism, Environmentalism, and Public Health in Edwardian England," *Victorian Studies* 34 (Winter 1991): 159–78, and W. F. Bynum, "Policing Hearts of Darkness: Aspects of the International Sanitary Conferences," *History and Philosophy of the Life Sciences* 15, no. 3 (1993): 421–34.

9. Quoted in Eyler, *Victorian Social Medicine,* 100.

10. See, for example, *Times* (London), 3 Jan. 1850.

11. W. Noel Hartley, *Water, Air, and Disinfectants* (London: Society for Promoting Christian Knowledge, [1877]), 89.

12. *Times* (London), 23 Aug. 1855, 5e.

13. *The Cheap Doctor: A Word about Fresh Air* (London: Ladies' National Association for the Diffusion of Sanitary Knowledge, [1859]), 7.

14. Balthazar Foster, "Colds and Coughs," in *Birmingham Health Lectures,* 1st ser. (Birmingham: Hudson, 1883), 34. For an interesting discussion of evolving attitudes toward nighttime air in the United States, see Peter C. Baldwin, "How Night Air Became Good Air, 1776–1930," *Environmental History* 8 (July 2003): 412–29.

15. Arthur Ransome, *Foul Air and Lung Disease* (Manchester: Manchester and Salford Sanitary Association, [ca. 1877]), 27.

16. Alfred Carpenter, "The First Principles of Sanitary Work," *Sanitary Record,* 15 Dec. 1879, 203–4, quotation on 204.

17. Quoted in Eyler, *Victorian Social Medicine,* 99. The most effective means of ventilating interiors, according to many Victorian experts, was a hearty fire. While this propelled large volumes of air up the chimney, it also filled the air outdoors with smoke. See Stephen Mosley, "Fresh Air and Foul: The Role of the Open Fireplace in Ventilating the British Home, 1837–1910," *Planning Perspectives* 18, no. 1 (2003): 1–21.

18. "A London Fog," *Leisure Hour,* 1 Dec. 1853, 772–74, quotation on 773.

19. "Observations in a London Fog," *Hogg's Instructor* 5 (1855): 53–55, quotation on 55.

20. "Frost, Fog, and Smoke," *Lancet,* 2 Jan. 1892, 40.

21. Robert Angus Smith, *On Some Invisible Agents of Health and Disease* (Manchester: Manchester and Salford Sanitary Association, 1878), 105.

22. See Mary J. Dobson, *Contours of Death and Disease in Early Modern England* (Cambridge: Cambridge University Press, 1997).

23. See Saul Jarcho, "A Cartographic and Literary Study of the Word *Malaria*," *Journal of the History of Medicine and the Allied Sciences* 25, no. 1 (1970): 31–39. For a nineteenth-century account, see Thomas Herbert Barker, *On Malaria and Miasmata, and Their Influence in the Production of Typhus and Typhoid Fevers, Cholera, and the Exanthemata* (London: John W. Davies, 1863).

24. J. White, *On Health, as Depending on the Condition of Air. . . .* (London: Hamilton, Adams, 1859), 3. See also Christopher Hoolihan, "Health and Travel in Nineteenth-Century Rome," *Journal of the History of Medicine and the Allied Sciences* 44, no. 4 (1989): 462–85.

25. George Rosen, *A History of Public Health,* expanded ed., introduction by Elizabeth Fee (Baltimore: Johns Hopkins University Press, 1993), 298–99.

26. "A London Fog," *Leisure Hour,* 774.

27. *Times* (London), 23 Aug. 1855, 5e.

28. Edward Headlam Greenhow, *On the Study of Epidemic Disease, as Illustrated by the Pestilences of London* (London: T. Richards, 1858), 18.

29. "A London Fog," *Leisure Hour,* 774.

30. This observation, though made of France, is equally applicable to mid-nineteenth-century Britain. See Alain Corbin, *The Foul and the Fragrant: Odor and the French Social Imagination,* trans. Miriam L. Kochan (Cambridge: Harvard University Press, 1986), 66.

31. Roger Cooter, "Anticontagionism and History's Medical Record," in *The Problem of Medical Knowledge: Examining the Social Construction of Medicine,* ed. Peter Wright and Andrew Treacher, 87–108 (Edinburgh: Edinburgh University Press, 1982). See also Ludmilla Jordanova, "The Social Construction of Medical Knowledge," *Social History of Medicine* 8 (Dec. 1995): 361–81.

32. Paul Slack, *The Impact of Plague in Tudor and Stuart England* (Oxford: Clarendon Press, 1985), 45, 205, 245, 249. Fires were also used for this purpose in North America; see Charles-Edward Amory Winslow, *The Conquest of Epidemic Disease* (Princeton: Princeton University Press, 1943), 196.

33. John Evelyn, *Fumifugium, or the Inconvenience of the Aer and Smoake of London Dissipated. . . .* (1661), repr. in *The Smoake of London: Two Prophecies,* ed. James P. Lodge (Elmsford, N.Y.: Maxwell Reprint, 1969), 14–15.

34. John Charles Atkinson, *Change of Air: Fallacies Regarding It* (London: John Ollivier, 1848), 26.

35. Robert Angus Smith, "On the Air of Towns," *Quarterly Journal of the Chemical Society* 11 (1859): 224.

36. "The Opening of the Smoke Abatement Exhibition at South Kensington," *Sanitary Record*, 15 Dec. 1881, 227–60, esp. 227.

37. Peter Spence, speaking from the floor at the meeting held on 14 Dec. 1858, *Proceedings of the Literary and Philosophical Society of Manchester* 1 (1857–60): 80. On the amalgam of science and industry that Spence, the Literary and Philosophical Society, and Manchester embodied during this period, see Robert H. Kargon, *Science in Victorian Manchester: Enterprise and Expertise* (Baltimore: Johns Hopkins University Press, 1977).

38. Peter Spence, *Coal, Smoke, and Sewage, Scientifically and Practically Considered* (Manchester, 1857), 18.

39. "Chemical and Physical Modifications," 147.

40. Manchester Steam Users' Association for the Prevention of Steam Boiler Explosions and for the Attainment of Economy in the Application of Steam, *A Sketch of the Foundation and of the Past Fifty Years' Activity* (Manchester: Taylor, Garnet, Evans, 1905).

41. Neil Arnott, *On the Smokeless Fire-Place, Chimney-Valves, and Other Means, Old and New, of Obtaining Healthful Warmth and Ventilation* (London: Longmans, Brown, Green, and Longmans, 1855), 3.

42. John P. Seddon, "The Warming of Houses," *Architect*, 14 Dec. 1872, 327–29.

43. Repr. in John Simon, *Public Health Reports*, ed. Edward Seaton, 2 vols. (London: Sanitary Institute of Great Britain, 1887), 1: 67–69. See also Royston Lambert, *Sir John Simon, 1816–1904, and English Social Administration* (London: MacGibbon and Kee, 1963).

Chapter 3

1. Cornelius B. Fox, *Sanitary Examinations of Water, Air, and Food* (London: J. and A. Churchill, 1878), 181–82.

2. Bill Luckin, " 'The Heart and Home of Horror': The Great London Fogs of the Late Nineteenth Century," *Social History* 28 (Jan. 2003): 31–48.

3. For a sampling of this vast literature, see John M. Eyler, *Victorian Social Medicine: The Ideas and Methods of William Farr* (Baltimore: Johns Hopkins University Press, 1979), 65; John Ashton and Howard Seymour, *The New Public Health: The Liverpool Experience* (Milton Keynes, England: Open University Press, 1988; repr. 1995), 17; Margaret Pelling, "Contagion/Germ Theory/Specificity," in *Companion Encyclopedia of the History of Medicine*, ed. W. F. Bynum

and Roy Porter, 1:309–34 (London: Routledge, 1993); W. F. Bynum, *Science and the Practice of Medicine in the Nineteenth Century* (Cambridge: Cambridge University Press, 1994); Mervyn Susser and Ezra Susser, "Choosing a Future for Epidemiology, I: Eras and Paradigms," *American Journal of Public Health* 86 (May 1996): 668–73.

4. See William Napier Shaw, "The Treatment of Smoke: A Sanitary Parallel," *Journal of the Sanitary Institute* 23, no. 3 (1902): 318–34.

5. Joe St. Loe Strachey to Thomas Coglan Horsfall, 16 Dec. 1898. Horsfall Collection, Letters (P–W), 374 (emphasis in original). Quoted by permission of Manchester Archives and Local Studies. On Horsfall's influence, see Michael Harrison, "Thomas Coglan Horsfall and 'the Example of Germany,'" *Planning Perspectives* 6, no. 3 (1991): 297–314.

6. "Meeting of the British Medical Association at Cork: Public Medicine Section," *Sanitary Record*, 15 Sept. 1879, 104–9. On the history of ideas about susceptibility to illness, see Christopher Hamlin, "Predisposing Causes and Public Health in Nineteenth-Century Medical Thought," *Social History of Medicine* 5 (Apr. 1992): 43–70.

7. Joseph Fayrer, "Inaugural Address," *Journal of the Sanitary Institute* 19 (Oct. 1898): 337–59, quotation on 359.

8. "Dust and Fog," *Engineering*, 7 Jan. 1881, 15.

9. "Fogs," *Lancet*, 23 Oct. 1880, 665–66.

10. W. R. E. Coles, "Smoke Abatement," *Transactions of the Sanitary Institute of Great Britain* 7 (1885): 218–28. Chaumont's words come from the discussion that accompanies Coles's paper; see 225.

11. "London Fogs," *Spectator*, 9 Jan. 1892, 45–46.

12. Richard Pritchard, "The Influence of Ventilation on the Type of the Disease," *Public Health* 15 (1902–3): 385–93, esp. 387.

13. W. Noel Hartley, *Water, Air, and Disinfectants* (London: Society for Promoting Christian Knowledge, [1877]), 89.

14. *Dictionary of Scientific Biography*, s.v. "Schönbein, Christian Friedrich."

15. J. C., "Ozone," *Once a Week*, 14 Jan. 1865, 94–96, quotation on 94.

16. John Makinson Fox, *Defective Drainage as a Cause of Disease*, Health Lectures for the People (Manchester: Manchester and Salford Sanitary Association, [1879]), 90.

17. John Angell, *Personal and Household Arrangements in Relation to Health* (Manchester and Salford Sanitary Association, [ca. 1878]), 19 (emphasis in original). On the history of this group, see P. A. Ryan, "Public Health and Voluntary Effort in Nineteenth-Century Manchester, with Particular Reference to the Manchester and Salford Sanitary Association," master's thesis, University of Manchester, [1974].

18. Arthur Ransome, *Foul Air and Lung Disease* (Manchester: Manchester and Salford Sanitary Association, [ca. 1877]), 6–7.

19. *Health and Meteorology of Manchester* 1, no. 1 (1860) (copy in Box M3C, Manchester Medical Collection, John Rylands University Library, Manchester).

20. Benjamin Ward Richardson, "On Ozone in Relation to Health and Disease," *Popular Science Review* 5 (1866): 29–40, quotations from 33–34, 40.

21. Henry Bollmann Condy, *Air and Water: Their Impurities and Purification* (London: John W. Davies, 1862), 13; Robert Angus Smith, *On Some Invisible Agents of Health and Disease* (Manchester: Manchester and Salford Sanitary Association, 1878); Arthur Ransome, "On the Distribution of Death and Disease," *Health Journal* 1 (Apr. 1884): 168–69.

22. Ransome, *Foul Air*, 12.

23. Frederic L. Holmes, "Elementary Analysis and the Origins of Physiological Chemistry," *Isis* 54, no. 1 (1963): 50–81, esp. 50, 55.

24. Edward Bascome, *Prophylaxis, or the Mode of Preventing Disease by a Due Appreciation of the Grand Elements of Vitality: Light, Air, and Water, with Observations on Intramural Burials* (London: S. Highley, 1849), 4.

25. Not everyone was quite so sanguine, however. See Christopher Hamlin, "Environmental Sensibility in Edinburgh, 1839–1840: The 'Fetid Irrigation' Controversy," *Journal of Urban History* 20 (May 1994): 311–39.

26. William H. Brock, *Justus von Liebig: The Chemical Gatekeeper* (Cambridge: Cambridge University Press, 1997), 203–4.

27. Benjamin Ward Richardson, "Health and Civilisation," *Journal of the Society of Arts,* 15 Oct. 1875, 948–54, quotation on 954.

28. Arthur Downes and Thomas P. Blunt, "The Influence of Light upon the Development of Bacteria," *Nature,* 12 July 1877, 218; idem, "Researches on the Effect of Light upon Bacteria and Other Organisms," *Proceedings of the Royal Society of London,* 6 Dec. 1877, 488–500; idem, "On the Influence of Light upon Protoplasm," *Proceedings of the Royal Society of London,* 19 Dec. 1878, 199–212.

29. W. H. Stone, "On Fog," *Popular Science Review,* n.s., 5, no. 17 (1881): 27–39, esp. 30.

30. "Fogs," *Lancet,* 23 Oct. 1880, 665–66. On the high incidence of fatal lung diseases, see G. Melvyn Howe, *People, Environment, Disease and Death: A Medical Geography of Britain through the Ages* (Cardiff: University of Wales Press, 1997), 175.

31. Roy Porter, *The Greatest Benefit to Mankind: A Medical History of Humanity* (New York: W. W. Norton, 1997), 555.

32. F. Rufenacht Walters, *Sanatoria for Consumptives: A Critical and Detailed Description,* 3rd ed. (London: Swan Sonnenschein, 1905), 6.

33. C. B. Underhill, "Recollections of Life at an Open-Air Sanatorium," *Good Words and Sunday Magazine* 44 (1903): 183–86. Similar ideas influenced domestic architecture. See Annmarie Adams, *Architecture in the Family Way: Doctors, Houses, and Women, 1870–1900* (Montreal and Kingston: McGill–Queen's University Press, 1996).

34. A. Wynter Blyth, "Ventilation," *Public Health* 14 (1901–2): 61–91, quotation on 61–62. On the role of the journal *Public Health*, see Margaret Pelling, "'Progress, Difficulties, Suggestions, and Reforms': *Public Health*, 1888–1974," *Public Health* 102, no. 3 (1988): 209–15.

35. Percy Alden, "Coal Smoke Abatement," *Contemporary Review* (Dec. 1922): 725–33, quotations on 726–27.

36. Harvey Littlejohn, *Report on the Causes and Prevention of Smoke from Manufacturing Chimneys* (1897), 24–25 (copy at Sheffield Local Studies Library). On the role of the "MOH," see Huw Francis, "Understanding Medical Officers of Health," *Public Health* 102 (1988): 545–53.

37. William Blake Richmond to Angie Acland, 24 Dec. 1896 (MS. Acland, d. 160, fol. 46, Bodleian Library Special Collections and Western Manuscripts, Oxford University).

38. *Times* (London), 30 Jan. 1880, 10f.

39. "The Fog in London," *Lancet*, 3 Jan. 1874, 28.

40. "The Fogs of London," *Nature*, 23 Dec. 1880, 165–66; Ministry of Health, *Mortality and Morbidity during the London Fog of December 1952* (London: HMSO, 1954), 60.

41. *Times* (London), 25 Dec. 1879; Robert H. Scott, "London Fogs," *Longman's Magazine* 9 (Apr. 1887): 607–14, esp. 607.

42. Charles Dickens, *Our Mutual Friend*, ed. Michael Cotsell (Oxford: Oxford University Press, 1989), 420.

43. Louis C. Parkes, "The Air and Water of London: Are They Deteriorating?" *Transactions of the Sanitary Institute of Great Britain* 13 (1892): 59–69, esp. 62.

44. Francis Albert Rollo Russell, "Haze, Fog, and Visibility," *Quarterly Journal of the Royal Meteorological Society* 23 (1897): 10–24, quotation on 19.

45. Robert H. Scott, "Fifteen Years' Fogs in the British Islands, 1876–1890," *Quarterly Journal of the Royal Meteorological Society* 19 (1893): 229–38, esp. 232–34. See also Francis Albert Rollo Russell, *The Atmosphere in Relation to Human Life and Health* (Washington, D.C.: Smithsonian Institution, 1896), 32–35.

46. Discussion of Alfred Carpenter, "London Fogs," *Journal of the Society of Arts*, 10 Dec. 1880, 48–60, quotation on 58.

47. Frederick J. Brodie, "On the Prevalence of Fog in London during the Years 1871 to 1890," *Quarterly Journal of the Royal Meteorological Society* 18 (1892): 40–45, quotation on 43.

48. Scott, "Fifteen Years' Fogs," 232–34.

49. Attempts to deny human culpability when disaster strikes are by no means limited to Britain. See Theodore Steinberg, *Acts of God: An Unnatural History of Natural Disaster in America* (New York: Oxford University Press, 2000), esp. xvii–xx, 151–52.

50. *Times* (London), 27 Dec. 1904, 11a.

Chapter 4

1. B. H. Thwaite, "London Fog: A Scheme to Abolish It," *National Review* 20 (Nov. 1892): 360–67.

2. Richard Price, *A Supplement to the Second Edition of the Treatise* (London: T. Cadell, 1772), 12–13.

3. Donald Worster, *Nature's Economy: A History of Ecological Ideas,* 2nd ed. (Cambridge: Cambridge University Press, 1994), x, 50.

4. Roy Porter, *The Greatest Benefit to Mankind: A Medical History of Humanity* (New York: W. W. Norton, 1997), 325.

5. S. Scott Alison, *An Inquiry into the Propagation of Contagious Poisons, by the Atmosphere; as also into the Nature and Effects of Vitiated Air, Its Forms and Sources, and Other Causes of Pestilence. . . .* (Edinburgh: Maclachan, Stewart, 1839), 100.

6. Benjamin Ward Richardson, *On the Poisons of the Spreading Diseases: Their Nature and Mode of Distribution* (London: John Churchill, 1867), 27. This is the text of a talk delivered before the Congress on the Sewage Question in Leamington, 25 Oct. 1866.

7. Clemens Winkler, "The Influence of the Combustion of Coal upon Our Atmosphere," *Open Court* 1 (May 1887): 197–99, quotation on 198–99.

8. C. M. Aikman, *Air, Water, and Disinfectants* (London: Society for Promoting Christian Knowledge, 1895), 22, 44.

9. Svante Arrhenius, "On the Influence of Carbonic Acid in the Air upon the Temperature of the Ground," *London, Edinburgh, and Dublin Philosophical Magazine,* 5th ser., 41 (Apr. 1896): 237–76. See also Julia Uppenbrink, "Arrhenius and Global Warming," *Science* 272, no. 5265 (1996): 1122; Henning Rodhe and Robert Charlson, eds., *The Legacy of Svante Arrhenius: Understanding the Greenhouse Effect* (Stockholm: Royal Swedish Academy of Sciences, 1998).

10. Arnold Zuckerman, "Disease and Ventilation in the Royal Navy: The Woodenship Years," *Eighteenth-Century Life* 11 (Nov. 1987): 77–89; William H. Brock, *The Chemical Tree: A History of Chemistry* (New York: W. W. Norton, 2000), 107.

11. Thomas Southwood Smith, *The Common Nature of Epidemics and Their Relation to Climate and Civilization; also Remarks on Contagion and Quarantine*, ed. T. Baker (London: N. Trübner, 1866), 12.

12. F. S. B. François de Chaumont, *The Habitation in Relation to Health* (London: Society for Promoting Christian Knowledge, 1879), 80.

13. Eric Stuart Bruce, "Town Fogs: Their Amelioration and Prevention," *Dublin Review* 114 (Jan.–Apr. 1894): 132–44, esp. 137–38. In contrast to carbon monoxide, which is highly toxic, carbon dioxide does not interfere with human respiration as long as oxygen is available.

14. "Proceedings of the Health Section," *Sanitary Record*, 20 Oct. 1879, 138–41, esp. 140–41. Further claims about the ability of eucalyptus and pine to counteract miasma by oxidizing the air occur in Charles T. Kingzett, *Nature's Hygiene: A Series of Essays on Popular Scientific Subjects. . . .* (London: Baillière, Tindall, and Cox, 1880), 111–13, 190–91.

15. Alfred Carpenter, *Preventive Medicine in Relation to the Public Health* (London: Simpkin, Marshall, 1877), 69. Concerns about the importance of balance between the body and the wider world of course predated the late nineteenth century. See John V. Pickstone, "Dearth, Dirt, and Fever Epidemics: Rewriting the History of British 'Public Health,' 1780–1850," in *Epidemics and Ideas: Essays in the Historical Perception of Pestilence*, ed. Terence Ranger and Paul Slack (Cambridge: Cambridge University Press, 1992), 129.

16. William Francis Cowper-Temple, First Baron Mount-Temple, "President's Address," *Report of the Proceedings of the Commons Preservation Society, 1882–83* (1883) (copy at Surrey History Centre, Woking).

17. Louis C. Parkes, "The Air and Water of London: Are They Deteriorating?" *Transactions of the Sanitary Institute of Great Britain* 13 (1892): 59–69, quotation on 62.

18. See, for example, Reginald Brabazon, "The London County Council and Open Spaces," *New Review* 7 (Dec. 1892): 701–7.

19. London Liberal and Reform Union, *London's Lungs* [1896] (copy at Surrey History Centre, Woking). See also Reginald Brabazon, "Lungs for Our Great Cities," *New Review* 2 (May 1890): 432–43.

20. Roy Church, with the assistance of Alan Hall and John Kanefsky, *The History of the British Coal Industry*, vol. 3, *1830–1913: Victorian Pre-Eminence* (Oxford: Clarendon Press, 1986), 20; Aikman, *Air, Water, and Disinfectants*, 44.

21. J. R. McNeill, *Something New under the Sun: An Environmental History of the Twentieth-Century World* (New York: W. W. Norton, 2000), 8.

22. Thwaite, "London Fog," 360.

23. James Burn Russell, *An Address Delivered at the Opening of the Section of Public Medicine, at the Annual Meeting of the British Medical Association in Sheffield, August 1876* (Glasgow: Robert Anderson, 1876), 16. Repr. in idem, *Public Health Administration in Glasgow: A Memorial Volume of the Writings of James Burn Russell,* ed. Archibald Kerr Chalmers (Glasgow: James Maclehose, 1905), 139.

24. Herbert Fletcher, *The Smoke Nuisance* (Manchester: Manchester and Salford Noxious Vapours Abatement Association, 1888), 155 (copy at British Library, London).

25. Robert Holland, *Air Pollution as Affecting Plant Life* (Manchester: Manchester and Salford Noxious Vapours Abatement Association, 1888), 121.

26. A. D. Webster, *Town Planting and the Trees, Shrubs, Herbaceous and Other Plants That Are Best Adapted for Resisting Smoke* (London: George Routledge, [1910]), 8, 67, 74, 191–92.

27. On eighteenth-century books that recommended smoke-resistant flowers for London, see Keith Thomas, *Man and the Natural World: A History of the Modern Sensibility* (New York: Pantheon, 1983), 235. In late nineteenth-century Glasgow the superintendent of parks offered similar advice in letters to local newspapers. See George Carruthers Thomson, *On Smoke Abatement with Reference to Steam Boiler Furnaces, Read before the Philosophical Society of Glasgow, 6th February 1895* (repr., Calcutta Boiler Commission, n.d.), 17; R. S. R. Fitter, *London's Natural History* (London: Collins, 1945), 86.

28. George Vivian Poore, "Light, Air, and Fog," *Transactions of the Sanitary Institute of Great Britain* 14 (1893): 13–41, esp. 32.

29. Douglas Strutt Galton, *Army Sanitation: A Course of Lectures Delivered at the School of Military Engineering, Chatham,* 2nd ed. (Chatham: Royal Engineers Institute, 1887), 28.

30. W. H. Stone, "On Fog," *Popular Science Review,* n.s., 5, no. 17 (1881): 27–39, quotation on 38.

31. John Collins, "Air and Ventilation," *Transactions of the Sanitary Institute of Great Britain* 6 (1884–85): 391–95, quotation on 392 (emphasis in original).

32. Robert Barr, "The Doom of London," *Idler* 2 (1892–93): 397–409; repr. in James P. Lodge, ed., *The Smoake of London: Two Prophecies* (Elmsford, N.Y.: Maxwell Reprint, 1969), 44–56, quotation on 48. On fog deaths in 1891, see "Casualties Due to Fog and Frost," *Lancet,* 2 Jan. 1892, 33; Ministry of Health, *Mortality and Morbidity during the London Fog of December 1952* (London: HMSO, 1954), 60.

33. Owen C. D. Ross, "A Cure for London Fogs," *Gentleman's Magazine* 274 (Mar. 1893): 232.

34. Edwin Chadwick, *Report on the Sanitary Condition of the Labouring Population of Great Britain,* ed. M. W. Flinn (1842; Edinburgh: Edinburgh University Press, 1965), 135–50.

35. Edwin Chadwick, "Ventilation with Air from Superior Layers in Place of Inferior Layers" (1886), 2 (Chadwick Papers, 35/18, University College London, Rare Books Room).

36. A fine reproduction of Cruikshank's drawing can be found in Kenneth Pearson, ed., *Drawn and Quartered: The World of the British Newspaper Cartoon, 1720–1970* (London: Times Newspapers, [1970]).

37. That Christianity is to blame for Western environmental problems was famously suggested in Lynn White Jr., "The Historical Roots of Our Ecologic Crisis," *Science* 155, no. 3767 (1967): 1203–7. For a more recent assessment, see Peter Harrison, "Subduing the Earth: Genesis 1, Early Modern Science, and the Exploitation of Nature," *Journal of Religion* 79, no. 1 (1999): 86–109.

38. "Mr. Gladstone on Sanitary Matters," *British Architect and Northern Engineer,* 10 Aug. 1877, 70–71. See also Agatha Ramm, "Gladstone's Religion," *Historical Journal* 28, no. 2 (1985): 327–40. On natural theology in nineteenth-century Britain, see Graeme Davison, "The City as a Natural System: Theories of Urban Society in Early Nineteenth-Century Britain," in *The Pursuit of Urban History,* ed. Derek Fraser and Anthony Sutcliffe, 349–70 (London: Edward Arnold, 1983) ; Christopher Hamlin, "Providence and Putrefaction: Victorian Sanitarians and the Natural Theology of Health and Disease," *Victorian Studies* 28 (Spring 1985): 381–411.

39. J. Carter Bell, *Noxious Vapours Which Pollute the Air* (Manchester: Manchester and Salford Noxious Vapours Abatement Association, [1888?]) (copy at British Library, London), 1, 11. Such language came easily to many late-nineteenth-century writers. In 1899, for example, a newspaper reported that polluted air of Birmingham made the parish of St. Laurence "resemble at times a veritable Gehenna." See *Daily Mail* (London), 16 May 1899.

40. Alfred Beaumont Maddock, *On Sydenham, Its Climate and Place, with Observations on the Efficacy of Pure Air, Especially When Combined with Intellectual and Physical Recreation in the Prevention and Treatment of Disease* (London: Simkin, Marshall, 1860), iv.

41. John Leigh, *Coal-Smoke: Report to the Health and Nuisance Committees of the Corporation of Manchester* (Manchester: John Heywood, 1883), 3–5.

42. William Morris, *The Earthly Paradise* (1868–70), quoted in G. Robert Stange, "The Frightened Poets," in *The Victorian City: Images and Realities,* ed.

H. J. Dyos and Michael Wolff, 2 vols. (London: Routledge and Kegan Paul, 1973), 2: 488.

43. Albert Wilson, "The Great Smoke-Cloud of the North of England and Its Influence on Plants," abstract in the *Report of the Seventieth Meeting of the British Association for the Advancement of Science Held in Bradford September 1900* (London: John Murray, 1900), 930–31.

44. *Times* (London), 25 Oct. 1881, 11c.

45. Nicholas Cooke, "On Melanism in Lepidoptera," *Entomologist* 10 (Apr. 1877): 95.

46. Jeremy Cherfas, "Clean Air Revives the Peppered Moth," *New Scientist*, 2 Jan. 1986, 17; L. Doncaster, "Collective Inquiry as to Progressive Melanism in Lepidoptera," *Entomologist's Record and Journal of Variation* 18 (1906): 165–68, 206–54, esp. 250.

47. Doncaster, "Collective Inquiry," 251.

48. Nicholas Cooke, "Melanism in Lepidoptera," *Entomologist* 10 (June 1877): 151–53; F. Buchanan White, "Melanism, &c., in Lepidoptera," *Entomologist* 10 (May 1877): 126–29.

49. Fitter, *London's Natural History,* 182–83.

50. The precise connection between air pollution and melanism remains the subject of controversy. Recent research raises doubts that the advantage of dark over light moths in smoky environments results from the former's ability to hide in sooty trees, and suggests that dark moths may possess a greater genetic resiliency to pollution. See D. R. Lees, E. R. Creed, and J. G. Duckett, "Atmospheric Pollution and Industrial Melanism," *Heredity* 30 (1973): 227–32; Cherfas, "Clean Air," 17.

51. Augustus Voelcker, "On the Injurious Effects of Smoke on Certain Building Stones and on Vegetation," *Journal of the Society of Arts,* 22 Jan. 1864, 146–53, quotation on 150.

52. Douglas Galton, *On Some Preventible Causes of Impurity in London Air* (London, 1880), 8, 10.

53. *Times* (London), 27 Sept. 1877, 9c; William Crookes, *The Wheat Problem, Based on Remarks Made in the Presidential Address to the British Association in Bristol in 1898* (London: John Murray, 1899), 4; B. R. Mitchell, *International Historical Statistics: Europe, 1750–1993,* 4th ed. (London: Macmillan Reference, 1998), 311, 403.

54. See "The Reign of Darkness," *Spectator,* 19 Jan. 1889, 85.

55. Poore, "Light, Air, and Fog," 33.

56. Arthur Ransome, *Foul Air and Lung Disease* (Manchester: Manchester and Salford Sanitary Association, [ca. 1877]), 28. On the feminization of nature

in Victorian discourse, see Barbara T. Gates, *Kindred Nature: Victorian and Edwardian Women Embrace the Living World* (Chicago: University of Chicago Press, 1998), 3.

Chapter 5

1. Frederic Harrison, "A Few Words about the Nineteenth Century," *Fortnightly Review* 37 (Apr. 1882): 411–26, quotation on 422.

2. H. A. Des Voeux, *Smoke Abatement* (Manchester: National Smoke Abatement Society, 1936), 4–5.

3. Quoted in *Manchester Guardian,* 3 Nov. 1876.

4. Richard Jefferies, *After London, or Wild England* (London: Cassell, 1886), 378–79.

5. Robert Holland, *Air Pollution as Affecting Plant Life* (Manchester: Manchester and Salford Noxious Vapours Abatement Association, 1888), 124.

6. "Sir William B. Richmond, R.A., K.C.B., on the Smoke Nuisance," *London Argus,* 14 Jan. 1899, 218.

7. William Booth, *In Darkest England, and the Way Out* (London: Salvation Army, 1890; repr. Montclair, N.J.: Patterson Smith, 1975), 47–48.

8. "Dr. Arnott on Smokeless Fires and Pure Air in Houses," *Chambers's Journal,* 15 Sept. 1855, 174–76, quotation on 175.

9. William Blasius, "Some Remarks on the Connection of Meteorology with Health," paper read on 17 Dec. 1875, *Proceedings of the American Philosophical Society* 14 (1875): 667–71, quotation on 667.

10. London County Council, *Smoke Nuisance in London: Report of the Chief Officer of the Public Control Department* (1904), 4 (copy at London Metropolitan Archives).

11. Donald Olsen, *The Growth of Victorian London* (New York: Holmes and Meier, 1976), 23–24, 283; K. S. Inglis, *Churches and the Working Classes in Victorian England* (London: Routledge and Kegan Paul, 1963), 146.

12. *Manchester Guardian,* 5 Dec. 1877, 7.

13. Holland, *Air Pollution,* 124–25.

14. Arthur Ransome, *Foul Air and Lung Disease* (Manchester: Manchester and Salford Sanitary Association, [ca. 1877]), 8.

15. Booth, *In Darkest England.* See also Gareth Stedman Jones, *Outcast London: A Study in the Relationship between Classes in Victorian Society* (Oxford: Clarendon Press, 1971; repr., New York: Pantheon, 1984).

16. George R. Sims, *How the Poor Live* (1883), repr. in *How the Poor Live, and Horrible London* (London: Chatto and Windus, 1889), 64, 106–7.

17. H. G. Wells, *The Time Machine* (1895), repr. in H. G. Wells, *The Time Machine, The Wonderful Visit, and Other Stories* (New York: Charles Scribner's Sons, 1924), 1:60–61, 63, 65.

18. Daniel Ellis, *Considerations Relative to Nuisance in Coal-Gas Works. . . .* (Edinburgh: John Anderson, 1828), 18–19.

19. Friedrich Engels, *The Condition of the Working Class in England in 1844* (1845), trans. W. O. Henderson and W. H. Chaloner (Stanford: Stanford University Press, 1968), 70.

20. *Report of the Commissioners Appointed to Inquire into the Several Matters Relating to Coal in the United Kingdom*, iii. [C. 435–2], H.C. 1871, xviii, 815. See also Alfred S. Harvey, "Our Coal Supply," *Macmillan's Magazine* 26 (Sept. 1872): 375–84, esp. 375.

21. "Notes," *Nature*, 2 Apr. 1885, 513.

22. Leonhard Sohncke, "The Problem of the Exhaustion of Coal," *Open Court* 4 (July 1890): 2375–76, 2389–92.

23. John W. Graham, *The Destruction of Daylight: A Study in the Smoke Problem* (London: George Allen, 1907), 134–35, 139.

24. Peter Brears, *Images of Leeds, 1850–1960* (Derby: Breedon Books, 1992), 65.

25. T. E. C. Leslie, "The Known and the Unknown in the Economic World," *Fortnightly Review* 31 (June 1879): 934–49, quotation on 934.

26. *Illustrated London News*, 3 Jan. 1880, 2.

27. Des Voeux, *Smoke Abatement*, 6.

28. Holland, *Air Pollution*, 122–23.

29. Francis Albert Rollo Russell, *The Atmosphere in Relation to Human Life and Health* (Washington, D.C.: Smithsonian Institution, 1896), 35.

30. "The Smoke Nuisance," *Builder*, 12 Aug. 1899, 143–45.

31. William Bousfield, "Smoke in the Manufacturing Districts," *Art Journal* (1882): 9–10, quotation on 9.

32. Meath inherited the title earl when his father died in 1887; prior to that he was known as Lord Brabazon. For more on his life, see John O. Springhall, "Lord Meath, Youth, and Empire," *Journal of Contemporary History* 5, no. 4 (1970): 97–111; F. H. A. Aalen, "Lord Meath, City Improvement, and Social Imperialism," *Planning Perspectives* 4, no. 2 (1989): 127–52.

33. Reginald Brabazon, "Decay of Bodily Strength in Towns," *Nineteenth Century* 21 (May 1887): 673–76, quotation on 676.

34. Edward Carpenter, "Sunshine and Coal," *Daily News* (London), 28 May 1921.

35. *Sheffield Daily Telegraph*, 30 May 1921.

36. Ibid., 29 June 1921.

37. Ibid., 21 June 1922.

38. *Times* (London), 17 May 1921, 5c.

39. Ibid., 5b.

40. Ibid., 18 May 1921, 4c.

41. Ibid., 23 Aug. 1921, 9e.

42. A reproduction of this poster appears in Mark Kishlansky, Patrick Geary, and Patricia O'Brien, *Civilization in the West,* 3rd ed. (New York: Longman, 1998), 931.

43. James Phillips Kay, *The Moral and Physical Condition of the Working Classes Employed in the Cotton Manufacture in Manchester* (1832), quoted in Gary S. Messinger, *Manchester in the Victorian Age: The Halfknown City* (Manchester: Manchester University Press, 1985), 42. Nicholas Coles, "Sinners in the Hands of an Angry Utilitarian: J. P. Kay(-Shuttleworth), *The Moral and Physical Condition of the Working Classes in Manchester* (1832)," *Bulletin of Research in the Humanities* 86, no. 4 (1983–85): 453–88. Upon his marriage in 1842, Kay changed his name to Kay-Shuttleworth.

44. Alexis de Tocqueville, *Journeys to England and Ireland* (1835), ed. J. P. Mayer, trans. George Lawrence and K. P. Mayer (New Brunswick, N.J.: Transaction Press, 1988), 107.

45. Yi-Fu Tuan, *Landscapes of Fear* (Minneapolis: University of Minnesota Press, 1979), 152.

46. Mark J. Bouman, "The 'Good Lamp is the Best Police' Metaphor and Ideologies of the Nineteenth-Century Urban Landscape," *American Studies* 32 (Spring 1991): 63–78.

47. William Frederick Pollock, "Smoke Prevention," *Nineteenth Century* 9 (Mar. 1881): 478–90, quotation on 480.

48. Stedman Jones, *Outcast London,* 291–92.

49. Francis Albert Rollo Russell, *London Fogs* (London: Edward Stanford, 1880), 31–33.

50. Frank Harris, "The Radical Programme, III: The Housing of the Poor in Towns," *Fortnightly Review* 40 (Oct. 1883): 587–600, quotation on 596.

51. *Lancet,* 15 Dec. 1883, 1050.

52. *Times* (London), 25 Dec. 1879.

53. William Morris, *Art and Socialism,* 2nd ed. (London: Leek Bijou Reprint, [1884]), 21.

54. "Sir William B. Richmond," *London Argus,* 218.

55. "The Smoke Nuisance," 143.

56. "The Fogs of London," cutting from unidentified newspaper, 21 Nov. 1882 (copy at Royal Academy of Arts, London) (RI/4/5/3); William Blake

Richmond, "The Black City," *Pall Mall Magazine* 23 (Apr. 1901): 462–73, quotation on 468.

57. *Times* (London), 13 Mar. 1882, 12b.

58. "Fogs of London" (RI/4/5/3).

59. *Times* (London), 17 Jan. 1901, 3e.

60. Coal Smoke Abatement Society, *Summary of Law Relating to Smoke Pollution* (London, 1901) (copy at National Society for Clean Air and Environmental Protection, Brighton).

61. See Martin A. Danahay, "Matter out of Place: The Politics of Pollution in Ruskin and Turner," *Clio* 21, no. 1 (1991): 61–77; Steven Z. Levine, *Monet and His Critics* (New York: Garland, 1976).

62. Quoted in John House, *Monet: Nature into Art* (New Haven: Yale University Press, 1986), 129, 222.

63. M. H. Dziewicki, "In Praise of London Fog," *Nineteenth Century* 26 (Dec. 1889): 1047–55, quotation on 1054–55.

64. Harrison, "A Few Words," 415.

65. John Ruskin, "The Lamp of Memory" (1849), in *The Seven Lamps of Architecture,* 2nd ed. (New York: John Wiley, 1890), 274, 301.

66. E. T. Cook, *The Life of John Ruskin,* 2nd ed., 2 vols. (London: George Allen, 1912), 1:470. See also Denis Cosgrove and John E. Thornes, "Of Truth of Clouds: John Ruskin and the Moral Order in Landscape," in *Humanistic Geography and Literature,* ed. Douglas C. D. Pocock, 20–46 (London: Croom Helm, 1981); Raymond E. Fitch, *The Poison Sky: Myth and Apocalypse in Ruskin* (Athens: Ohio University Press, 1982).

67. For an unfavorable contemporary assessment of Ruskin's address, see "Mr. Ruskin in the Clouds," *Builder,* 9 Feb. 1884, 190–91.

68. James Burn Russell, *Public Health Administration in Glasgow: A Memorial Volume of the Writings of James Burn Russell,* ed. Archibald Kerr Chalmers (Glasgow: James Maclehose, 1905), 278.

69. Quoted in Cook, *Ruskin,* 1:337.

70. Society for the Protection of Ancient Buildings, *Annual Report* (1882), 23–24.

71. National Trust, "Report of the Provisional Council" (1895), 1–2, National Trust, London; Sally Mitchell, ed., *Victorian Britain: An Encyclopedia* (New York: Garland, 1988), 185.

72. David Lowenthal, *The Past Is a Foreign Country* (Cambridge: Cambridge University Press, 1985, repr. 1990), 96.

73. Ruskin, "Lamp of Memory," 285.

74. Charles Dellheim, *The Face of the Past: The Preservation of the Medieval Inheritance in Victorian England* (Cambridge: Cambridge University Press,

1982), 130. See also the perceptive comments of Patrick Wright, *On Living in an Old Country: The National Past in Contemporary Britain* (London: Verso, 1985), esp. 48–56.

75. Jane Fawcett, "A Restoration Tragedy: Cathedrals in the Eighteenth and Nineteenth Centuries," in *The Future of the Past: Attitudes to Conservation, 1174–1974*, ed. Jane Fawcett (London: Thames and Hudson, 1976), 113.

76. John Lubbock, "On the Preservation of Our Ancient National Monuments," *Nineteenth Century* 1 (Apr. 1877): 257–69, quotation on 266.

77. "Copies of Circulars. . . . " (1908), LC/TC/Reports, F139, Leeds District Archives.

78. Dellheim, *Face of the Past,* 92–112.

79. Robert Angus Smith, "On the Air of Towns," *Quarterly Journal of the Chemical Society* 11 (1859): 196–235, esp. 209, 232.

80. Stephanie M. Blackden, "The Development of Public Health Administration in Glasgow, 1842–1872" (Ph.D. thesis, Edinburgh University, 1976), 283.

81. "Condition of Our Chief Towns—Sheffield," *Builder,* 21 Sept. 1861, 641.

82. Edward Carpenter, "Coal and Wet Weather: An Object Lesson from Sheffield," *Daily News* (London), 21 June 1921.

83. "Report on the Decay of Stone at Westminster," *Builder,* 5 Oct. 1861, 677–80, esp. 679.

84. Augustus Voelcker, "On the Injurious Effects of Smoke on Certain Building Stones and on Vegetation," *Journal of the Society of Arts,* 22 Jan. 1864, 146–53, quotations on 146–47, 150.

85. "The Smoke Nuisance," *Builder,* 31 Dec. 1881, 835.

86. H. T., "Dr. Mohr on 'The Atmosphere and Our Obelisk,'" *Builder,* 9 Mar. 1878, 251.

87. A.R.I.B.A., "The Obelisk," *Builder,* 23 Feb. 1878, 198; "Cleopatra's Needle," *British Architect and Northern Engineer,* 23 Mar. 1877, 174–75; "The Obelisk," *Builder,* 12 Oct. 1878, 1074; Edward Carpenter, "The Smoke-Plague and Its Remedy," *Macmillan's Magazine* 62 (July 1890): 204–13, quotation on 206.

88. Ruskin, "Lamp of Memory," 284–85. For another example of the prevalence of this sentiment in Victorian culture, see "Public Monuments," *Edinburgh Review* 115 (Apr. 1862): 548.

89. Sidney Colvin, "Restoration and Anti-Restoration," *Nineteenth Century* 2 (Oct. 1877): 456–61, quotation on 456.

90. Edward Carpenter, "Beautiful Sheffield," manuscript of a speech delivered in April 1910, C. F. Sixsmith Edward Carpenter Collection, Eng 1171/4/1. Reproduced by courtesy of the University Librarian and Director, John Rylands University Library, University of Manchester. On Carpenter, see Sheila

Rowbotham and Jeffrey Weeks, eds., *Socialism and the New Life: The Personal and Sexual Politics of Edward Carpenter and Havelock Ellis* (London: Pluto Press, 1977).

91. Octavia Hill, "The Future of Our Commons," *Fortnightly Review* 28 (Nov. 1877): 631–41, quotation on 641.

92. John Simon, *Public Health Reports,* ed. Edward Seaton, 2 vols. (London: Sanitary Institute of Great Britain, 1887), 1:66.

93. F. A. R. Russell, *London Fogs,* 31–33.

94. *Times* (London), 21 Aug 1885, 8e. On Barnett, see Bentley B. Gilbert, *The Evolution of National Insurance in Great Britain: The Origins of the Welfare State* (London: Michael Joseph, 1966), 40, 43, 52; P. W. J. Bartrip, *Mirror of Medicine: A History of the British Medical Journal* (Oxford: Clarendon Press, 1990), 64; Henrietta O. Barnett, *Canon Barnett: His Life, Work, and Friends,* 2 vols. (London: John Murray, 1919), 1:71, 134, 185; 2:46–47.

95. Morris, *Art and Socialism,* 42.

96. "The Smoke-Abatement Exhibition," *Lancet,* 17 Dec. 1881, 1068–69, quotation on 1068.

97. Holland, *Air Pollution as Affecting Plant Life,* 112.

98. Graham, *Destruction of Daylight,* 13.

99. For a perceptive discussion of working-class interest in outdoor recreation and the back-to-nature movement during the early twentieth century, see Raphael Samuel, *Island Stories: Unravelling Britain,* ed. Alison Light, with Sally Alexander and Gareth Stedman Jones (London: Verso, 1998).

100. Benjamin Stott, *Songs for the Millions and Other Poems* (Middleton: W. Horsman, 1843), x.

101. Ebenezer Howard, *To-morrow: A Peaceful Path to Real Reform* (London: Swan Sonnenschein, 1898).

Chapter 6

1. Alfred Carpenter, *Preventive Medicine in Relation to the Public Health* (London: Simpkin, Marshall, 1877), 63.

2. H. G. Wells, *The Time Machine* (1895), repr. in H. G. Wells, *The Time Machine, The Wonderful Visit, and Other Stories* (New York: Charles Scribner's Sons, 1924), 1:60–65.

3. W. H. Michael, "The Law in Relation to Sanitary Progress," paper read before the Sanitary Institute of Great Britain on 11 Feb. 1881, *Sanitary Record,* 15 Feb. 1881, 281–85, quotation on 282.

4. *Glasgow Herald,* 30 Oct. 1890.

224 | Notes to Pages 69-72

5. Review of Degeneration: A Chapter in Darwinism, by E. Ray Lankester, London Quarterly Review 56, no. 112 (1881): 353–66, esp. 353.

6. John Edward Morgan, The Danger of Deterioration of Race from the Too Rapid Increase of Great Cities (London: Longmans, Green, and Co., 1866), 15.

7. John Milner Fothergill, The Town Dweller: His Needs and His Wants (London: Lewis, 1889), 4, 16–17.

8. Madison Grant, The Passing of the Great Race, or the Racial Basis of European History (New York: Charles Scribner's Sons, 1916), 186.

9. See Bernard Semmel, Imperialism and Social Reform: English Social-Imperial Thought, 1895–1914 (Cambridge: Harvard University Press, 1960); Bentley B. Gilbert, "Health and Politics: The British Physical Deterioration Report of 1904," Bulletin of the History of Medicine 39, no. 2 (Mar.–Apr. 1965): 143–53; G. R. Searle, The Quest for National Efficiency: A Study in British Politics and Political Thought, 1899–1914 (Oxford: Basil Blackwell, 1971); Greta Jones, Social Darwinism and English Thought: The Interaction between Biological and Social Theory (Brighton: Harvester, 1980); Richard A. Soloway, Demography and Degeneration: Eugenics and the Declining Birthrate in Twentieth-Century Britain (Chapel Hill: University of North Carolina Press, 1990); Marouf Arif Hasian, The Rhetoric of Eugenics in Anglo-American Thought (Athens: University of Georgia Press, 1996); and Mike Hawkins, Social Darwinism in European and American Thought, 1860–1945: Nature as a Model and Nature as a Threat (Cambridge: Cambridge University Press, 1997).

10. Robert Jones, "Physical and Mental Degeneration," Journal of the Society of Arts, 10 Mar. 1904, 327–43, quotations from 328, 330, 337, 338, 340.

11. Morgan, Danger, 11, 50.

12. Lyon Playfair, "Pure Air, Pure Water, and Pure Soil," Builder, 10 Oct. 1874, 846.

13. Edwin Chadwick asserted in the 1840s that army volunteers from Manchester were less fit than in previous decades. He also claimed that fathers were often considerably taller than their adult sons, another sign of progressive deterioration. See Edwin Chadwick, Report on the Sanitary Condition of the Labouring Population of Great Britain (1842), ed. M. W. Flinn (Edinburgh: Edinburgh University Press, 1965), 251.

14. Quoted in James Fraser, "Opening Address," Transactions of the National Association for the Promotion of Social Science, Manchester meeting, 1879, 1–31, quotation from 30.

15. Reginald Brabazon, "Decay of Bodily Strength in Towns," Nineteenth Century 21 (May 1887): 673–76.

16. Thomas E. Jordan, *The Degeneracy Crisis and Victorian Youth* (Albany: State University of New York Press, 1993), esp. 141; and William Greenslade, review of *The Degeneracy Crisis and Victorian Youth,* by Thomas E. Jordan, *Victorian Studies* 38, no. 2 (1995): 273–74.

17. *Times* (London), 26 Dec. 1903, 8e.

18. Anthony S. Wohl, *Endangered Lives: Public Health in Victorian Britain* (Cambridge: Harvard University Press, 1983); Dorothy Porter, "'Enemies of the Race': Biologism, Environmentalism, and Public Health in Edwardian England," *Victorian Studies* 34 (Winter 1991): 159–78.

19. Inter-Departmental Committee on Physical Deterioration, *Minutes of Evidence,* Cd. 2210 (1904), par. 2310, 4226.

20. William Nicholson, *Smoke Abatement: A Manual for the Use of Manufacturers, Inspectors, Medical Officers of Health, Engineers, and Others* (London: Charles Griffin, 1905), 2.

21. The report was summarized in *Times* (London), 29 July 1904, 7e.

22. Inter-Departmental Committee on Physical Deterioration, *Report,* Cd. 2175 (1904), 85–86.

23. Semmel, *Imperialism and Social Reform,* 44–45.

24. *Times* (London), 4 Jan. 1904, 5e.

25. Karl Pearson, *National Life from the Standpoint of Science* (1905), quoted in Semmel, *Imperialism and Social Reform,* 41, 48–49.

26. Ethel M. Elderton, *The Relative Strength of Nurture and Nature* (London: Dulau, 1909), 33.

27. Gerry Kearns, "Biology, Class and the Urban Penalty," in *Urbanising Britain: Essays on Class and Community in the Nineteenth Century,* ed. Gerry Kearns and W. J. Withers, 12–30 (Cambridge: Cambridge University Press, 1991), esp. 24.

28. Fothergill, *Town Dweller,* 112–13.

29. While Fothergill ascribed degeneration to miscegenation, Engels suggested that the English urban poor might degenerate by adopting habits and customs brought by Irish immigrants. See Friedrich Engels, *The Condition of the Working Class in England in 1844* (1845), trans. W. O. Henderson and W. H. Chaloner (Stanford: Stanford University Press, 1968), 107. I thank Michael de Nie for bringing this to my attention.

30. Sidney Webb, *The Decline in the Birth Rate* (1907), quoted in Semmel, *Imperialism and Social Reform,* 51.

31. Grant, *Passing of the Great Race,* 186–87.

32. "The Sanitary Inspectors' Association: Conference at Lincoln," *Builder,* 12 Aug. 1899, 155–57, quotation on 156–57.

33. Thomas Glover Lyon, "The Air Supply of London," *Journal of Preventive Medicine* 13 (1905): 685–89, esp. 685.

34. David Fraser-Harris, *National Degeneration, Being the Annual Public Lecture on the Laws of Health Delivered at the Midland Institute, 17 Sept. 1909* (Birmingham: Cornish Brothers, 1909), 24. For an insightful analysis of this phenomenon in another context, see Andrew Hurley, *Environmental Inequalities: Class, Race, and Industrial Pollution in Gary, Indiana, 1945–1980* (Chapel Hill: University of North Carolina Press, 1995).

35. Review of *Degeneration*, 359.

36. Fraser-Harris, *National Degeneration*, 30.

37. Ibid., 32.

38. G. R. Searle, *Eugenics and Politics in Britain, 1900–1914* (Leyden: Noordhoff, 1976), 108–11.

39. "The Crime of Being Inefficient," *Nation* (London), 25 May 1912, 275.

40. Ibid., 275–77; ibid., 15 June 1912, 391.

41. Dorothy Porter, "Eugenics and the Sterilization Debate in Sweden and Britain before World War II," *Scandinavian Journal of History* 24, no. 2 (1999): 145–62. On contemporaneous developments in the United States, see Angela Gugliotta, "'Dr. Sharp with His Little Knife': Therapeutic and Punitive Origins of Eugenic Vasectomy: Indiana, 1892–1921," *Journal of the History of Medicine and Allied Sciences* 53, no. 4 (1998): 371–406.

42. Morgan, *Danger*, 5.

43. W. R. E. Coles, "Smoke Abatement," *Transactions of the Sanitary Institute of Great Britain* 5 (1883): 335–45, quotation on 336.

44. Reginald Brabazon, "Health and Physique of Our City Populations," *Nineteenth Century* 10 (July 1881): 81–83, 85.

45. Reginald Brabazon, "Great Cities and Social Reform, I," *Nineteenth Century* 14 (Nov. 1883): 798–809, esp. 802.

46. Nancy Stepan, "Biological Degeneration: Races and Proper Places," in *Degeneration: The Dark Side of Progress*, ed. J. Edward Chamberlin and Sander L. Gilman (New York: Columbia University Press, 1985), 98–99. See also Nancy Stepan, *The Idea of Race in Science: Great Britain, 1800–1960* (London: Macmillan, 1982); idem, *"The Hour of Eugenics": Race, Gender, and Nation in Latin America* (Ithaca: Cornell University Press, 1991).

47. For a discussion of links between environmental and social order, see Roger Cooter, "The Power of the Body: The Early Nineteenth Century," in *Natural Order: Historical Studies of Scientific Culture*, ed. Barry Barnes and Steven Shapin, 73–92 (Beverly Hills: Sage, 1979), and Bill Luckin, *Pollution and Control: A Social History of the Thames in the Nineteenth Century* (Bristol: Adam Hilger, 1986), esp. 20–30.

48. Daniel Pick, *Faces of Degeneration: A European Disorder, c. 1848–c. 1918* (Cambridge: Cambridge University Press, 1989), 218.

49. *Times* (London), 14 Oct. 1884, 3c.

50. Brabazon, "Decay," 675.

51. Paul Weindling, *Health, Race, and German Politics between National Unification and Nazism, 1870–1945* (Cambridge: Cambridge University Press, 1989).

Chapter 7

1. H. A. Des Voeux, *Smoke Abatement* (Manchester: National Smoke Abatement Society, 1936), 5.

2. Brian Harrison, *Peaceable Kingdom: Stability and Change in Modern Britain* (Oxford: Clarendon Press, 1982), 398; Jane Lewis, *Women and Social Action in Victorian and Edwardian England* (Stanford: Stanford University Press, 1991).

3. Ronald K. Huch, "The National Association for the Promotion of Social Science: Its Contribution to Victorian Health Reform, 1857–1886," *Albion* 17 (Fall 1985): 279–300; F. Prochaska, "A Mother's Country: Mothers' Meetings and Family Welfare in Britain, 1850–1950," *History* 74 (Oct. 1989): 379–99, esp. 391; Kathleen E. McCrone, "The National Association for the Promotion of Social Science and the Advancement of Victorian Women," *Atlantis* 8 (Fall 1982): 44–66; David Ward, "Victorian Cities: How Modern?" *Journal of Historical Geography* 1 (1975): 135–51, esp. 146.

4. Eileen Janes Yeo, "Social Motherhood and the Sexual Communion of Labour in British Social Science, 1850–1950," *Women's History Review* 1, no. 1 (1992): 63–87, esp. 80.

5. Birmingham City Council, Minutes (1880–82), 63–64 (Birmingham City Archives).

6. Birmingham Ladies' Association for Useful Work, *Annual Report* 5 (1880): 8 (emphasis in original) (copy at Local Studies and History Service, Birmingham).

7. Ladies' Sanitary Association, *The Twenty-Third Annual Report* (London, 1881), esp. 8.

8. "Chemical and Physical Modifications of the Atmosphere Consequent on Habitation," *Chemical News,* 15 Mar. 1862, 146.

9. See, for example, J. Carter Bell, *Noxious Vapours Which Pollute the Air* (Manchester: Manchester and Salford Noxious Vapours Abatement Association, [1888?]).

10. *Times* (London), 5 Nov. 1880.

11. *Report from the Select Committee on Coal,* H.C. 1873 (313), x, 1, p. 316.

12. William Napier Shaw and John Switzer Owens, *The Smoke Problem of Great Cities* (London: Constable, 1925), 41.

13. Coal Smoke Abatement Society, Minutes, 4 Mar. 1904, 37 (National Society for Clean Air and Environmental Protection, Brighton).

14. Francis Albert Rollo Russell, *London Fogs* (London: Edward Stanford, 1880), 11.

15. Sidney Barwise, "The Abatement of the Smoke Nuisance," *Sanitary Record,* 15 June 1891, 609.

16. "Manchester: Smoke," *Lancet,* 28 Apr. 1894, 1099–1100, quotation on 1099.

17. B. Morley Fletcher, *Report [of the] Committee for Testing Smoke-Preventing Appliances* (Manchester, 1896), 22.

18. Charles Estcourt, *Why the Air of Manchester Is So Impure* (Manchester: Manchester and Salford Noxious Vapours Abatement Association, [1887]), 41.

19. Lawson Tait, *Gas as Fuel* (Birmingham: Hudson, n.d.), 13.

20. "The Smoke Abatement Exhibition," *Nature,* 5 Jan. 1882, 219–21, esp. 219.

21. *Encyclopædia Britannica,* 9th ed., s.v. "smoke abatement."

22. Thomas Coglan Horsfall, *The Nuisance of Smoke from Domestic Fires, and Methods of Abating It* (Manchester: Manchester and Salford Noxious Vapours Abatement Association, [1893?]), 2–4.

23. "The Sanitary Institute Congress at Leeds," *Builder,* 18 Sept. 1897, 222–23, quotation on 223.

24. "The Coal Smoke Abatement Society and Domestic Grates," *Lancet,* 10 May 1902, 1332–35, quotation on 1332.

25. Martin H. Daly, "The Smoke Nuisance," *Brewers' Journal,* 15 Feb. 1909 124–25. I owe this reference to Jonathan Reinarz.

26. William B. Smith, "Should the Domestic Smoke Nuisance Be Any Longer Tolerated?" in Coal Smoke Abatement Society, *Papers Read at the Smoke Abatement Conferences, March 26, 27, and 28, 1912, with Discussions* (London: Coal Smoke Abatement Society, 1912), 62–67. In addition to serving as an elected member of Glasgow's municipal government, Smith led the local branch of the Smoke Abatement League of Great Britain.

27. Inter-Departmental Committee on Physical Deterioration, *Minutes of Evidence,* Cd. 2210 (1904), par. 6457.

28. Donald J. M. Smith, "The Eaton Hall Connection," *Transport History* 9, no. 3 (1978): 256–59. See also Gervas Huxley, *Victorian Duke: The Life of Hugh Lupus Grosvenor, First Duke of Westminster* (London: Oxford University Press, 1967).

29. Peter Stallybrass and Allon White, *The Politics and Poetics of Transgression* (Ithaca: Cornell University Press, 1986), 145, 191, 202.

30. R. Arthur Arnold, *Sanitary Reform: "Water Supply" and "Pure Air"* (London: Bradbury, Evans, 1866), 29.

31. Birmingham City Council, Sanitary Committee, Minutes (1874–76), minute 4824 (Birmingham City Archives).

32. *Annual Report Made to the Urban Sanitary Authority of the Borough of Leeds for the Year 1893*, table 11, and *Annual Report Made to the Urban Sanitary Authority of the Borough of Leeds for the Year 1898*, table 11. Copies at the Wellcome Library for the History and Understanding of Medicine, London.

33. Manchester and Salford Noxious Vapours Abatement Association, Minutes, 23 Sept. 1881 (Manchester Archives and Local Studies).

34. *How to Subdue Smoke: Being Popular Information on Various Practicable Means; with Comparative Results, and a Few Words about Fuel*, 3rd ed. (London: Effingham Wilson and Edward Stanford, 1854), 11.

35. "Correspondence," *Chemical News*, 28 July 1860, 82–83, quotation from 82.

36. William Nicholson, *Practical Smoke Prevention* (London: Sanitary Publishing, 1902), 30.

37. Walter Hepworth Collins, "The Alleged Danger to Public Health, Arising from Effluvium Nuisance from Gas Works," *Transactions of the Sanitary Institute of Great Britain* 11 (1890): 112–19, esp. 118.

38. Review of *London Smoke and Fog*, by Frederick Edwards, *Chemical News*, 7 Jan. 1881, 10–11.

39. Nicholson, *Practical Smoke Prevention*, viii, 27–29.

40. Peter W. Barlow, *Smoke Abatement* (London: Vacher, 1882), 8.

41. Antony Taylor, "'Commons-Stealers,' 'Land-Grabbers,' and 'Jerry-Builders': Space, Popular Radicalism, and the Politics of Public Access in London, 1848–1880," *International Review of Social History* 40, no. 3 (1995): 383–407.

42. See Charles Dellheim, *The Face of the Past: The Preservation of the Medieval Inheritance in Victorian England* (Cambridge: Cambridge University Press, 1982).

43. See S. Martin Gaskell, "Gardens for the Working Class: Victorian Practical Pleasure," *Victorian Studies* 23 (Summer 1980): 479–501; Howard LeRoy Malchow, "Public Gardens and Social Action in Late Victorian London," *Victorian Studies* 29 (Autumn 1985): 97–124.

44. For examples of its interests in these areas, see John Angell, "Address on the Paris Electrical and London Smoke-Abatement Exhibitions," paper read

on 24 Jan. 1882, *Report and Proceedings of the Manchester Field-Naturalists and Archaeologists' Society for the Year 1882*, 55–60; William H. Bailey, "Open Spaces and Recreational Grounds," *Report and Proceedings of the Manchester Field-Naturalists and Archaeologists' Society for the Year 1883*, 56–66; Manchester Field-Naturalists and Archaeologists' Society, *The Atmosphere of Manchester: Preliminary Report of the Air Analysis Committee* (1891).

45. Paul Readman, "Landscape Preservation, 'Advertising Disfigurement,' and English National Identity c. 1890–1914," *Rural History* 12, no. 1 (2001): 61–83.

46. On the National Trust, see John Gaze, *Figures in a Landscape: A History of the National Trust* (n.p.: Barrie and Jenkins, in association with the National Trust, 1988); Graham Murphy, *Founders of the National Trust* (London: Croom Helm, 1987); Jennifer Jenkins and Patrick James, *From Acorn to Oak Tree: The Growth of the National Trust, 1895–1994* (London: Macmillan, 1994).

47. See Stephen Mosley, *The Chimney of the World: A History of Smoke Pollution in Victorian and Edwardian Manchester* (Cambridge: White Horse Press, 2001), esp. 146–67.

48. Lawrence Goldman, "Debate: Reply," *Past and Present* 121 (Nov. 1988): 214–19, quotation on 218. See also idem, "The Social Science Association, 1857–1886: A Context for Mid-Victorian Liberalism," *English Historical Review* 101 (1986): 95–134; John Ranlett, "'Checking Nature's Desecration': Late-Victorian Environmental Organization," *Victorian Studies* 26 (Winter 1983): 197–222; Howard LeRoy Malchow, *Agitators and Promoters in the Age of Gladstone and Disraeli: A Biographical Dictionary of the Leaders of British Pressure Groups Founded between 1865 and 1886* (New York: Garland, 1983).

49. For examples of works that demonstrate this connection, see Carlos Flick, "The Movement for Smoke Abatement in 19th-Century Britain," *Technology and Culture* 21 (Jan. 1980): 29–50; Peter Brimblecombe, *The Big Smoke: A History of Air Pollution in London since Medieval Times* (London: Methuen, 1987); David Evans, *A History of Nature Conservation in Britain* (London: Routledge, 1992).

50. *Times* (London), 31 Mar. 1905.

51. See *Dictionary of National Biography*, s.v. "Hart, Ernest Abraham"; George K. Behlmer, "Ernest Hart and the Social Thrust of Victorian Medicine," *British Medical Journal*, 3 Oct. 1990, 711–13; P. W. J. Bartrip, *Mirror of Medicine: A History of the British Medical Journal* (Oxford: Clarendon Press, 1990), esp. 63–92.

52. Christopher Charlton, "The National Health Society Almanack, 1883," *Local Population Studies* 32 (Spring 1984): 54–57.

53. National Health Society, *Seventh Annual Report* (1880), 13.

54. "The National Health Society," *Sanitary Record,* 9 Feb. 1877, 89; "Clergymen as Sanitarians," *Sanitary Record,* 15 May 1881, 423–24; National Health Society, *Seventh Annual Report* (1880), 1–2, 4, 5–8.

55. Although much has been written about Hill, the secondary literature contains relatively little discussion of her involvement in smoke abatement. The major sources on Hill are *Dictionary of National Biography,* s.v. "Hill, Octavia"; C. Edmund Maurice, ed., *Life of Octavia Hill, as Told in Her Letters* (London: Macmillan, 1913); Enid Moberly Bell, *Octavia Hill: A Biography,* 2nd ed. (London: Constable, 1943); William Thomson Hill, *Octavia Hill: Pioneer of the National Trust and Housing Reformer* (London: Hutchinson, 1956); Nancy Boyd, *Three Victorian Women Who Changed Their World: Josephine Butler, Octavia Hill, Florence Nightingale* (New York: Oxford University Press, 1982); Murphy, *Founders of the National Trust,* esp. 45–51; Gaze, *Figures,* esp. 12–22; Gillian Darley, *Octavia Hill* (London: Constable, 1990). Most of what has been published about Hill's anti-smoke activism is contained in two short articles: Arnold Marsh, "Octavia Hill and Clean Air: A Smoke Abatement Pioneer," *Society of Housing Managers Quarterly Journal* 5 (Oct. 1962): 21–23; John Ranlett, "The Smoke Abatement Exhibition of 1881," *History Today* 31 (Nov. 1981): 10–13.

56. On Hill's and others' attempts to provide such housing, see J. N. Tarn, *Five Per Cent Philanthropy: An Account of Housing in Urban Areas between 1840 and 1914* (London: Cambridge University Press, 1973).

57. Bell, *Octavia Hill,* 150–53.

58. Bell, *Octavia Hill,* 150–53; Jan Marsh, *Back to the Land: The Pastoral Impulse in England from 1880 to 1914* (London: Quartet Books, 1982), 39; Darley, *Octavia Hill,* 179.

59. Octavia Hill, "More Air for London," *Nineteenth Century* 23 (Feb. 1888): 181–88, quotation on 183.

60. Hill, *Octavia Hill,* 107.

61. Bell, *Octavia Hill,* 170.

62. Henrietta O. Barnett, *Canon Barnett: His Life, Work, and Friends,* 2 vols. (London: John Murray, 1919), 1:71, 134; Bartrip, *Mirror of Medicine,* 64; Murphy, *Founders of the National Trust,* 58.

63. Maurice, *Life of Octavia Hill,* 261.

64. Miranda Hill to Mrs. Edmund Maurice, 27 Nov. 1874, in Maurice, *Life of Octavia Hill,* 311–19 (emphasis in original).

65. Barnett, *Canon Barnett,* 1:252–55; Maurice, *Life of Octavia Hill,* 417; *Times* (London), 6 Oct. 1930, 16; *Times* (London), 28 Feb. 1935, 13.

66. Octavia Hill to Caroline Southwood Hill, 24 May 1880, in Maurice, *Life of Octavia Hill,* 437.

67. Bell, *Octavia Hill,* 170.

68. National Health Society, *Eighth Annual Report* (1881), 15–18; Ernest Hart, "Report on the Abatement of Smoke and the Proceedings Adopted by the Smoke Abatement Committee of the National Health and Kyrle Societies," *Sanitary Record,* 15 Dec. 1881, 217–19; "Smoke-Abatement Exhibition," *Nature,* 8 Dec. 1881, 121–22, esp. 121.

69. *Times* (London), 25 Oct. 1880, 10e.

70. Ibid., 7 Dec. 1880, 12b.

71. "Fogs," *Lancet,* 23 Oct. 1880, 665–66.

72. "The Fog Question," *Chemical News,* 31 Dec. 1880, 327–28, quotation on 327.

73. Bell, *Octavia Hill,* 170.

74. Douglas Strutt Galton, *On Some Preventible Causes of Impurity in London Air* (London, 1880), 23.

75. Ibid., 15.

76. W. R. E. Coles, "Smoke Abatement," *Transactions of the Sanitary Institute of Great Britain* 7 (1885): 218–28. Hart's words appear on 224 as part of the discussion that followed Coles's paper.

77. Edward Frankland, "Smoke Abatement," *Nature,* 1 Mar. 1883, 407–8, esp. 408.

78. *Times* (London), 16 Jan. 1883, 3f.

79. Ibid., 12 Nov. 1880.

80. Ibid., 26 Nov. 1880.

81. Ibid., 13 Mar. 1882, 12b.

82. "The Opening of the Smoke Abatement Exhibition at South Kensington," *Sanitary Record,* 15 Dec. 1881, 227; *Manchester Guardian,* 1 Dec. 1881, 5a; "The Smoke Abatement Exhibition," *Nature,* 5 Jan. 1882, 219.

83. "The International Exhibition of Smoke-Preventing Appliances," *Builder,* 3 Dec. 1881, 693–707, esp. 693.

84. "Smoke Abatement Exhibition at South Kensington," *Sanitary Record,* 15 Feb. 1882, 339–40.

85. "The International Exhibition of Smoke-Preventing Appliances," *Builder,* 10 Dec. 1881, 739.

86. "The Smoke Abatement Exhibition," *Nature,* 5 Jan. 1882, 219.

87. "Smokeless Heat," *Builder,* 10 Dec. 1881, 715.

88. "Further Cogitations at the Smoke Abatement Exhibition," *Builder,* 17 Dec. 1881, 745–46.

89. "Smoke-Abating Appliances," *Lancet,* 3 Dec. 1881, 970.

90. "The Smoke-Abatement Exhibition," *Lancet,* 17 Dec. 1881, 1068–69, quotation on 1068.

91. Birmingham City Council, Health Committee, Minutes (1880–82), 348 (Birmingham City Archives).

92. "Smoke Abatement Exhibition at South Kensington," *Sanitary Record,* 15 Feb. 1882, 339–40.

93. Smoke Abatement Committee, *Report of the Smoke Abatement Committee, 1882* (London: Smith, Elder, 1883), 4.

94. *Times* (London), 13 Mar. 1882, 12b.

95. *Manchester Guardian,* 18 Mar. 1882, 13a; "The Manchester Smoke Abatement Exhibition," *Sanitary Record,* 15 Apr. 1882, 424–25, and 15 May 1882, 475–77.

96. The *Exhibition Review,* which was edited by Fred Scott, appeared in six installments between 5 April and 4 May 1882 (copy at National Society for Clean Air and Environmental Protection, Brighton).

97. "The Smoke Abatement Institute," *Nature,* 17 Aug. 1882, 371.

98. National Smoke Abatement Institution, *Smoke Abatement: Report of Council of National Smoke Abatement Institution, Submitted at a Public Meeting at the Mansion House, July 16th, 1884* (London, 1884), 10. For a comparison of smoke-abatement activism in Britain and the United States during this period, and for a discussion of interactions between reformers in the two countries, see David Stradling and Peter Thorsheim, "The Smoke of Great Cities: British and American Efforts to Control Air Pollution, 1860–1914," *Environmental History* 4 (Jan. 1999): 6–31. Daniel T. Rodgers analyzes European influences on other aspects of American social reform in his recent work, *Atlantic Crossings: Social Politics in a Progressive Age* (Cambridge: Belknap Press of Harvard University Press, 1998).

99. Alfred Carpenter to Edwin Chadwick, 24 Dec. 1883, Chadwick Papers, 448/18, University College London, Rare Books Room.

100. National Health Society, *Annual Report, 1884,* 19, 27.

101. *Report from the Select Committee of the House of Lords on Smoke Nuisance Abatement (Metropolis) Bill [H.L.],* pp. 34, 40 (313), H.C. 1887, xii, 265.

102. "The National Smoke Abatement Institution," *Builder,* 28 July 1883, 132.

103. National Smoke Abatement Institution, *Smoke Abatement,* 13; "Smoke Abatement," *Nature,* 19 July 1883, 279; *Report from the Select Committee of the House of Lords on Smoke,* xii.

104. Eric Ashby and Mary Anderson, *The Politics of Clean Air* (Oxford: Clarendon Press, 1981), 60–63.

105. George Percy, "Coal and Smoke," *Quarterly Review* 119 (Apr. 1866): 435–72; "The Ventilation of the House of Commons," *Sanitary Record,* 16 Feb. 1877, 97–99; "The Fog in Parliament," *Lancet,* 5 Jan. 1889, 35; *Times* (London), 14 July 1899, 10b.

106. Thomas Fletcher, *Flame and Smoke* (Manchester: Manchester and Salford Noxious Vapours Abatement Association, 1889), 4.

107. B. Morley Fletcher, *Report,* n.p.

108. "Smoke Abatement," *Engineering,* 7 Aug. 1896, 178–79, quotation on 178.

109. "Smoke Abatement," *Industries,* 7 Nov. 1890, 454.

110. B. Morley Fletcher, *Report,* 6–10.

111. Birmingham City Council, Health Committee, Minutes (1889–90), 334–35 (Birmingham City Archives).

112. Edward Carpenter, "Sunshine and Coal," *Daily News* (London), 28 May 1921.

113. Geoffrey Wall, "Public Response to Air Pollution in Sheffield, England" (preliminary paper prepared for the Man and Environment Commission, 22nd International Geographical Conference, Calgary, Alberta, Canada, July 1972), 5 (copy at Sheffield Local Studies Library).

114. S. O. Addy and W. T. Pike, *Sheffield at the Opening of the Twentieth Century: Contemporary Biographies* (Brighton: W. T. Pike, n.d.), 127.

115. Harvey Littlejohn, *Report on the Causes and Prevention of Smoke from Manufacturing Chimneys* (1897), 26, 38 (copy at Sheffield Local Studies Library).

116. Inter-Departmental Committee on Physical Deterioration, *Minutes of Evidence,* par. 6455, 4234.

117. William Graham, *Smoke Abatement in Lancashire* (Manchester: J. E. Cornish, 1896), 14, 16.

118. Manchester and Salford Sanitary Association, *Annual Report* (1895), 33–34; ibid. (1896), 35–49. Copies at Manchester Medical Collection, John Rylands University Library, Manchester.

119. "The Smoke Nuisance," *Builder,* 12 Aug. 1899, 143–45, quotation from 143.

120. Ibid., 143.

121. F. W. Oliver, "On the Effects of Urban Fog upon Cultivated Plants," *Journal of the Royal Horticultural Society* 13 (1891): 139–51, quotation on 150.

122. *Times* (London), 26 Jan. 1892.

123. Ibid., 30 Dec. 1904, 9a.

124. "Sir William B. Richmond, R.A., K.C.B., on the Smoke Nuisance," *London Argus,* 14 Jan. 1899, 217–18; Coal Smoke Abatement Society, Minutes, 15 Jan. 1904, 11 (National Society for Clean Air and Environmental Protection, Brighton).

125. Des Voeux, *Smoke Abatement,* 6.

126. *South London Press,* 23 Mar. 1901.

127. "Tests of Domestic Open Fire-Grates," *Lancet,* 10 May 1902, 1342.

128. Lawrence W. Chubb, *Smoke Abatement* (1912), 9 (copy at National Society for Clean Air and Environmental Protection, Brighton).

129. H. A. Des Voeux, *Ten Years' Work of Smoke Abatement in London* ([London: Coal Smoke Abatement Society, 1909]), 3–6.

130. Julius B. Cohen, "A Record of the Work of the Leeds Smoke Abatement Society," in *Addresses, Papers, and Discussions at the Conference on Smoke Abatement, Held in London, December 12–15th, 1905* (London: Royal Sanitary Institute, 1906), 27–29; *Times* (London), 14 Dec. 1905, 15d.

131. *Reports on the Laws in Force in Certain Foreign Counties in Regard to the Emission of Smoke from Chimneys,* Cd. 2347, H.C. 1905, lxxxv, 1.

132. Coal Smoke Abatement Society, Minutes, 5 Dec. 1898 (National Society for Clean Air and Environmental Protection, Brighton).

133. *Times* (London), 7 June 1883, 5f.

134. *Times* (London), 9 Jan. 1901, 12b, and 26 Feb. 1903, 12f.

135. "Sir William B. Richmond," *London Argus,* 217.

136. Lancet Analytical Sanitary Commission, "Report on Perfect Combustion and Smoke Prevention," *Lancet,* 21 Mar. 1891, 682–85; idem, "Second Report on Perfect Combustion and Smoke Prevention," *Lancet,* 5 Mar. 1892, 548–50; idem, "III—Gaseous Fuel, Gas-Heating, and Gas-Cooking Appliances," *Lancet,* 25 Nov. 1893, 1326–36.

137. Herbert Fletcher, *The Smoke Nuisance* (Manchester: Manchester and Salford Noxious Vapours Abatement Association, 1888), 156.

138. William Charles Popplewell, *The Prevention of Smoke, Combined with the Economical Combustion of Fuel* (London: Scott, Greenwood, 1901), xviii.

139. "Smoke," *Engineer,* 9 Dec. 1898, 569–70.

140. "Editorial Notes," *Industries and Iron,* 20 Jan. 1899, 41.

141. Reginald Brabazon, *Physical Training in Relation to the Needs of a Town Population* [1902], 5.

142. John W. Graham, *The Destruction of Daylight: A Study in the Smoke Problem* (London: George Allen, 1907), 2–3, 55–56, 144–45.

143. Smoke Abatement League of Great Britain, *First Annual Report* (1909–10), 4 (National Society for Clean Air and Environmental Protection, Brighton).

144. William Nicholson, "Smoke Abatement from the Inspector's Point of View," in *Papers Read at the Smoke Abatement Conferences,* 76–80.

145. John B. C. Kershaw, "Notes on Recent Progress in the Campaign against Black Smoke in This Country," in *Papers Read at the Smoke Abatement Conferences,* 68–75.

146. A. E. Haffner, "The Control of Atmospheric Pollution by Gasworks," address to the London and Southern Section of the Institution of Gas Engineers, 21 Apr. 1959 (copy at National Gas Archive, Warrington); F. Reynolds, *The Law Relating to Air Pollution* (Brighton: National Society for Clean Air, n.d.), 21.

Chapter 8

1. "Fogs," *Lancet,* 23 Oct. 1880, 665.

2. Fred Scott, *The Administration of the Law Relating to Smoke* [1895?], 4 (copy at National Society for Clean Air and Environmental Protection, Brighton).

3. "Pure Air and Noxious Vapours," *Builder,* 26 June 1880, 785.

4. Eric Ashby and Mary Anderson, *The Politics of Clean Air* (Oxford: Clarendon Press, 1981), 2, 4–6. See also Joel Franklin Brenner, "Nuisance Law and the Industrial Revolution," *Journal of Legal Studies* 3 (June 1974): 403–33.

5. Ibid., 16–18.

6. *Times* (London), 13 Nov. 1883, 2f.

7. See *Times* (London), 25 Dec. 1880.

8. *Architect,* 26 Apr. 1884, 262; Ashby and Anderson, *Politics of Clean Air,* 55; *Times* (London), 13 Nov. 1883, 2f; "Smoke Nuisance Abatement," *Lancet,* 8 Oct. 1887, 722–23.

9. Peter Brimblecombe, "Long Term Trends in London Fog," *Science of the Total Environment* 22, no. 1 (1981): 19–29, esp. 23.

10. See *Report of the Select Committee of the House of Lords on Injury from Noxious Vapours,* H.L. 1862 (486), xiv, 1; A. E. Dingle, " 'The Monster Nuisance of All': Landowners, Alkali Manufacturers, and Air Pollution, 1828–64," *Economic History Review,* 2nd ser., 35, no. 4 (1982): 529–48; Roy M. MacLeod, "The Alkali Acts Administration, 1863–84: The Emergence of the Civil Scientist," *Victorian Studies* 9 (Dec. 1965): 83–112; Ashby and Anderson, *Politics of Clean Air,* esp. 32–33; Richard Hawes, "The Control of Alkali Pollution in St. Helens, 1862–1890," *Environment and History* 1, no. 2 (1995): 159–71.

11. George Carruthers Thomson, *On Smoke Abatement with Reference to Steam Boiler Furnaces, Read before the Philosophical Society of Glasgow, 6th February 1895* (repr. Calcutta Boiler Commission, n.d.), 18.

12. See Alfred Evans Fletcher, *Present Legislation on Noxious Vapours, with Suggested Improvements* (Manchester: Manchester and Salford Noxious Vapours Abatement Association, 1888), 84.

13. Birmingham City Council, Health Committee, Minutes (1880–82), 284 (Birmingham City Archives).

14. Ashby and Anderson, *Politics of Clean Air,* 54–55, 60.

15. William Nicholson, *Practical Smoke Prevention* (London: Sanitary Publishing, 1902), 13–14.

16. See William Wallace, "The Air of Towns," in *Lectures on the Theory and General Prevention and Control of Infectious Diseases, and on Water Supply, Sewage Disposal, and Food,* by J. B. Russell and William Wallace (Glasgow: Robert Anderson, 1879), 133.

17. Ashby and Anderson, *Politics of Clean Air,* 73.

18. See J. Carter Bell, *Noxious Vapours Which Pollute the Air* (Manchester: Manchester and Salford Noxious Vapours Abatement Association, [1888?]), 10.

19. Ashby and Anderson, *Politics of Clean Air,* 84. The London Government Act of 1899 transferred the regulation of smoke in London to 27 newly created municipal boroughs and the Westminster City Council. The City of London retained its special privileges. See Ben Weinreb and Christopher Hibbert, eds., *The London Encyclopædia* (New York: St. Martin's Press, 1983), s.v. "vestries."

20. Joseph Hurst, *English Law Relating to the Emission of Smoke from Chimneys* (London: Coal Smoke Abatement Society, [ca. 1912]), 9–12 (copy at National Society for Clean Air and Environmental Protection, Brighton). Separate legislation existed for Scotland and for Ireland. See Anthony S. Wohl, *Endangered Lives: Public Health in Victorian Britain* (Cambridge: Harvard University Press, 1983), esp. 199–202.

21. Ashby and Anderson, *Politics of Clean Air,* 87.

22. London County Council, *Smoke Nuisance in London: Report of the Chief Officer of the Public Control Department* (1904), 9 (copy at London Metropolitan Archives).

23. John Davis, *Reforming London: The London Government Problem, 1855–1900* (Oxford: Clarendon Press, 1988), 132.

24. William Blake Richmond, "The Black City," *Pall Mall Magazine* 23 (Apr. 1901): 462–73, quotation on 463.

25. London County Council, *Smoke Nuisance,* 2–3.

26. *Times* (London), 10 Sept. 1898, 13e.

27. Birmingham City Council, Borough Inspection Committee, Minutes (1865–67), minutes 1799, 1815, 1832 (Birmingham City Archives).

28. Birmingham City Council, Borough Inspection Committee, Minutes (1867–71), minute 2474 (Birmingham City Archives).

29. Birmingham City Council, Sanitary Committee, Minutes (1871–74), minutes 3738, 3756 (Birmingham City Archives). On Cadbury's efforts to prevent pollution, see Maurice H. Bailey, "A Birmingham Clean Air Pioneer: The Influence of John Cadbury," *Smokeless Air* 102 (Summer 1957): 260–62.

30. Alfred Hill, *Report on the Health of the Borough of Birmingham* (1876), 16 and table vi (Local Studies and History Service, Birmingham).

31. Birmingham City Council, Sanitary Committee, Minutes (1874–76), minutes 4603, 4845, 4800, 4913 (Birmingham City Archives).

32. Birmingham City Council, Health Committee, Minutes (1880–82), 262 (Birmingham City Archives).

33. Ibid., 256.

34. Alfred Hill, *Report on the Health of the Borough of Birmingham* (1882), 29 (Local Studies and History Service, Birmingham).

35. Birmingham City Council, Health Committee, Minutes (1882–84), 318 (Birmingham City Archives).

36. Ibid., 346.

37. Ibid., 482.

38. Birmingham City Council, Health Committee, Minutes (1880–82), 181 (Birmingham City Archives).

39. Birmingham City Council, Health Committee, Minutes (1882–84), 192–93 (Birmingham City Archives).

40.Ibid., 210, 482.

41. Alfred Carpenter, "London Fogs," *Journal of the Society of Arts*, 10 Dec. 1880, 48–60, esp. 56.

42. W. R. E. Coles, "Smoke Abatement," *Transactions of the Sanitary Institute of Great Britain* 7 (1885): 218–28. Scott's words appear on 225–26 as part of a discussion that followed Coles's paper.

43. "Notes of the Sanitary Congress at York," *Builder*, 2 Oct. 1886, 499–502, esp. 500–1.

44. Herbert Fletcher, *The Smoke Nuisance* (Manchester: Manchester and Salford Noxious Vapours Abatement Association, 1888), 158, 161.

45. Harvey Littlejohn, *Report on the Causes and Prevention of Smoke from Manufacturing Chimneys* (1897), 12 (copy at Sheffield Local Studies Library).

46. "Dust and Smoke in Daily Life," *Lancet*, 14 Apr. 1906, 1054.

47. Inter-Departmental Committee on Physical Deterioration, *Minutes of Evidence*, Cd. 2210, H.C. 1904, par. 5404.

48. "The Smoke Nuisance," *Builder*, 12 Aug. 1899, 143–45, quotation from 143.

49. Review of *The Sanitary Inspector's Hand-Book*, by Albert Taylor, *Builder*, 7 July 1900, 10.

50. "Sanitary Inspectors' Association: Sixteenth Annual Dinner," *Builder*, 11 Feb. 1899, 147–48, quotations on 147.

51. Peter Fyfe, *Report of the Sanitary Department, Glasgow, for 1900*, 13 (copy at Glasgow University Archive Services).

52. William Nicholson, *Smoke Abatement: A Manual for the Use of Manufacturers, Inspectors, Medical Officers of Health, Engineers, and Others* (London: Charles Griffin, 1905), 6.

53. William Chambers, "A London Fog," *Chambers's Journal*, 4 Dec. 1880, 769–71, quotation from 770.

54. Francis Albert Rollo Russell, *Smoke in Relation to Fogs in London* (London, 1888), 14–15.

55. Birmingham City Council, Health Committee, Minutes (1880–82), 256 (Birmingham City Archives).

56. Birmingham City Council, Health Committee, Minutes (1889–90), 61, 76 (Birmingham City Archives).

57. Thomas J. Bass, *Tragedies of Life: A Fragment of To-day* (Birmingham, [ca. 1903]), 67.

58. Ibid., 66.

59. Ibid., 26–27.

60. Ibid., 25.

61. Birmingham City Council, Health Committee, Minutes (1901–2), 303–4 (Birmingham City Archives).

62. A. K. Chalmers, *The Health of Glasgow, 1818–1925: An Outline* (Glasgow: Corporation [of the City], 1930), 450.

63. Littlejohn, *Report*, 39.

64. Fyfe, *Report*, 28–29.

65. Inter-Departmental Committee on Physical Deterioration, *Minutes of Evidence*, par. 4234.

66. London County Council, *Smoke Nuisance*, 3.

67. Julius B. Cohen, "A Record of the Work of the Leeds Smoke Abatement Society," in *Addresses, Papers, and Discussions at the Conference on Smoke Abatement, Held in London, December 12–15th, 1905* (London: Royal Sanitary Institute, 1906), 28.

68. W. R. Hornby Steer, *The Law of Smoke Nuisances* (London: National Smoke Abatement Society, 1938), 14–15.

69. Birmingham City Council, Health Committee, Minutes (1878–80), 115, 127–28 (Birmingham City Archives).

70. "Smoke Abatement," *Engineering*, 7 Aug. 1896, 178–79, quotation on 178.

71. Medical Officer of Health, *Report on the Sanitary Condition of Leeds for the Year 1900*, n.p. (copy at Local Studies Library, Leeds).

72. Littlejohn, *Report*, 13–14, 44; Edward Carpenter, "The Smoke-Plague and Its Remedy," *Macmillan's Magazine* 62 (July 1890): 204–13, quotation from 204.

73. *Journal of the Society of Arts* 3 (1854–55): 146.

74. Birmingham City Council, Health Committee, Minutes (1878–80), minute 7300 (Birmingham City Archives).

75. Birmingham City Council, Health Committee, Minutes (1882–84), 374 (Birmingham City Archives).

76. Ibid., 211, 224.

77. Susan D. Pennybacker, *A Vision for the Future, 1889–1914: Labour, Everyday Life, and the LCC Experiment* (London: Routledge, 1995), 115–16.

78. *Times* (London), 27 Mar. 1900, 11f.

79. Gwilym Gibbon and Reginald W. Bell, *History of the London County Council, 1889–1939* (London: Macmillan, 1939), 560.

80. Littlejohn, *Report*, 27.

81. Theodore Thomson, "Report of Inspectors Department for Month Ending February 29, 1888" (MD 3444/2, Archives, Sheffield City Libraries).

82. "Fighting the Smoke Nuisance: Factory Smoke at Sheffield—I," *Country Life*, 8 Jan. 1921, 33–34.

83. Louis Ascher, "Coal Smoke Abatement in England," *Journal of the Royal Sanitary Institute* 28 (Mar. 1907): 88–93, quotation on 92.

84. *Encyclopædia Britannica*, 9th ed., s.v. "smoke abatement." For a discussion of an early scrubber, see "Elliott's Smoke Annihilator," *Sanitary Record*, 16 Feb. 1891, 429–30.

85. Thomas Coglan Horsfall, *The Nuisance of Smoke from Domestic Fires, and Methods of Abating It* (Manchester: Manchester and Salford Noxious Vapours Abatement Association, [1893?]), 5 (copy at British Library, London).

86. Julius B. Cohen and Arthur G. Ruston, *Smoke: A Study of Town Air* (London: Edward Arnold, 1912), 14–16.

87. *Glasgow Herald*, 16 Oct. 1909.

88. William Napier Shaw and John Switzer Owens, *The Smoke Problem of Great Cities* (London: Constable, 1925), vi, viii; Peter Brimblecombe, *The Big Smoke: A History of Air Pollution in London since Medieval Times* (London: Methuen, 1987), 103, 108; Albert Parker, "Air Pollution from the Use of Fuels," in *Industrial Air Pollution Handbook*, ed. Albert Parker (London: McGraw-Hill, 1978), 9.

89. Chalmers, *Health of Glasgow*, 456–57.

90. Alexander MacGregor, *Public Health in Glasgow, 1905–1946* (Edinburgh: E. and S. Livingstone, 1967), 78–79.

91. Department of Scientific and Industrial Research, *The Investigation of Atmospheric Pollution* 31 (1960), 9, discussed in T. J. Chandler, "Climate and the Built-Up Area," in *Greater London*, ed. J. T. Coppock and Hugh C. Prince (London: Faber and Faber, 1964), 43.

92. *Evening Standard* (London), 9 Oct. 1957.

93. Glasgow Sanitary Department, *Report on the Air of Glasgow . . . July 1877* (copy at Mitchell Library, Glasgow).

94. "The Air of Glasgow," *Sanitary Record,* 15 Jan. 1880, 261.

95. Manchester Field-Naturalists and Archaeologists' Society, *The Atmosphere of Manchester: Preliminary Report of the Air Analysis Committee* (1891); Julius B. Cohen and Arthur G. Ruston, *Smoke: A Study of Town Air* (London: Edward Arnold, 1912), 64–65.

96. Cited in William R. Smith, *The Laboratory Text-Book of Public Health* (London: Henry Renshaw, 1896), 149.

97. "The Analysis of the Air of Large Cities," *Industries,* 23 Jan. 1891, 91; *Yorkshire Evening News,* 21 Feb. 1924, 11 Dec. 1924; "Julius Berend Cohen," *Obituary Notices of Fellows of the Royal Society, 1932–35* 1 (1935): 503–13; Brimblecombe, *Big Smoke,* 149.

98. Ashby and Anderson, *Politics of Clean Air,* 56.

99. Peter Brimblecombe, "London Air Pollution, 1500–1900," *Atmospheric Environment* 11, no. 12 (1977): 1157–62, esp. 1161.

100. Ashby and Anderson, *Politics of Clean Air,* 82.

101. Frederick J. Brodie, "Decrease of Fog in London during Recent Years," *Quarterly Journal of the Royal Meteorological Society* 31 (1905): 15–28, esp. 16.

102. Brimblecombe, *Big Smoke,* 112; P. J. Waller, *Town, City, and Nation: England, 1850–1914* (Oxford: Oxford University Press, 1983), 30.

103. *Nature,* 14 May 1914, 274.

104. See Margaret White Fishenden, *The Efficiency of Low Temperature Coke in Domestic Appliances* (London: HMSO, 1921); Roy M. MacLeod and E. Kay Andrews, "The Origins of the D.S.I.R.: Reflections on Ideas and Men, 1915–1916," *Public Administration* 48 (Spring 1970): 23–48.

105. Ministry of Health, *Interim Report of the Committee on Smoke and Noxious Vapours Abatement* [Cmd. 755], H.C. 1920, xxv, 253.

106. Percy Alden, "Coal Smoke Abatement," *Contemporary Review,* Dec. 1922, 725–33, quotation on 732.

107. John B. C. Kershaw, "Smoke Abatement and the New Bill," *Fortnightly Review* 120 (Dec. 1923): 963–64.

108. *Times* (London), 8 Oct. 1929, 13b; Steer, *Law of Smoke Nuisances,* 4.

Chapter 9

1. W. E. Ayrton, "Electricity as a Motive Power," *Electrician,* 20 Sept. 1879, 215, quoted in Ido Yavetz, "A Victorian Thunderstorm: Lightning Protection

and Technological Pessimism in the Nineteenth Century," in *Technology, Pessimism, and Postmodernism,* ed. Yaron Ezrahi, Everett Mendelsohn, and Howard Segal (Dordrecht: Kluwer Academic Publishers, 1994), 67.

2. "The Domestic Smoke Nuisance," *Engineering,* 24 Nov. 1922, 653.

3. An earlier version of this chapter was published as "The Paradox of Smokeless Fuels: Gas, Coke, and the Environment in Britain, 1813–1949," *Environment and History* 8, no. 4 (2002): 381–401.

4. *Builder,* 25 Oct. 1884, 550.

5. Douglas Strutt Galton, *On Some Preventible Causes of Impurity in London Air* (London, 1880), 12–13.

6. Peter Spence, *Coal, Smoke, and Sewage, Scientifically and Practically Considered* (Manchester, 1857), 22, 28; idem, "Smoke and Fog," *Chemical News,* 25 Mar. 1881, 142–43.

7. William Geoffrey Goodacre, *A Scheme for Drawing Off and Disposing of All Waste Gases, Impure Air, &c.* (n.p., [1891]) (copy at National Society for Clean Air and Environmental Protection, Brighton).

8. A. Wynter Blyth, "Ventilation," *Public Health* 14 (1901–2): 61–91, esp. 62.

9. "The Power of the Future," *Electrician* 3 (13 Sept. 1879), quoted in W. P. Jolly, *Sir Oliver Lodge* (London: Constable, 1974), 43–44. Jolly attributes this anonymous article to Lodge, who later served on the council of the Coal Smoke Abatement Society.

10. "The Smoke Nuisance," *Builder,* 12 Aug. 1899, 143–45, quotation from 144.

11. "The Greenwich Observatory," *Lancet,* 16 June 1906, 1704.

12. *Times* (London), 10 Feb. 1930, 14d.

13. *Times* (London), 29 Oct. 1880, 6c.

14. B. R. Mitchell, *British Historical Statistics* (Cambridge: Cambridge University Press, 1988), 259.

15. My ideas about pollution displacement have been influenced by Martin V. Melosi, "Cities, Technical Systems and the Environment," *Environmental History Review* 14 (Spring-Summer 1990): 45–64; John T. Cumbler, "Whatever Happened to Industrial Waste? Reform, Compromise, and Science in Nineteenth Century Southern New England," *Journal of Social History* 29 (Fall 1995): 149–71; Joel A. Tarr, *The Search for the Ultimate Sink: Urban Pollution in Historical Perspective* (Akron, Ohio: University of Akron Press, 1996); Bill Luckin, "Pollution in the City," in *The Cambridge Urban History of Britain,* vol. 3, *1840–1950,* ed. Martin Daunton, 207–28 (Cambridge: Cambridge University Press, 2000).

16. Christine Meisner Rosen and Christopher C. Sellers, "The Nature of the Firm: Towards an Ecocultural History of Business," *Business History Review*

73, no. 4 (1999): 577–600, quotation from 584; Jeffrey K. Stine and Joel A. Tarr, "At the Intersection of Histories: Technology and the Environment," *Technology and Culture* 39 (Oct. 1998): 601–40, quotation from 621. Important works on the history of manufactured gas in Britain include M. E. Falkus, "The British Gas Industry before 1850," *Economic History Review,* 2d ser., 20, no. 3 (1967): 494–508; M. S. Cotterill, "The Development of Scottish Gas Technology, 1817–1914: Inspiration and Motivation," *Industrial Archaeology Review* 5, no. 1 (1980–81): 19–40; Trevor I. Williams, *A History of the British Gas Industry* (Oxford: Oxford University Press, 1981); Hugh Barty-King, *New Flame: How Gas Changed the Commercial, Domestic and Industrial Life of Britain between 1813 and 1984* (Tavistock, Devon: Graphmitre, 1984); J. F. Wilson, *Lighting the Town: A Study of Management in the North West Gas Industry, 1805–1880* (London: P. Chapman, 1991); Mary Mills, "The Early Gas Industry and Its Residual Products in East London," Ph.D. thesis, Open University, 1995. For an insightful analysis of the environmental consequences of coke production in the United States, and of the shifting burden of pollution from one medium to another, see Joel A. Tarr, "Searching for a 'Sink' for an Industrial Waste: Iron-Making Fuels and the Environment," *Environmental History Review* 18 (Spring 1994): 9–34 (republished in Tarr, *Search for the Ultimate Sink,* 385–411).

17. Eric Ashby and Mary Anderson, *The Politics of Clean Air* (Oxford: Clarendon Press, 1981), 86. See also Carlos Flick, "The Movement for Smoke Abatement in 19th-Century Britain," *Technology and Culture* 21 (Jan. 1980): 29–50; Peter Brimblecombe, *The Big Smoke: A History of Air Pollution in London since Medieval Times* (London: Methuen, 1987), esp. 112; Stephen Mosley, *The Chimney of the World: A History of Smoke Pollution in Victorian and Edwardian Manchester* (Cambridge: White Horse Press, 2001).

18. Williams, *A History of the British Gas Industry,* 9–10; Falkus, "British Gas Industry," 500, 504; Mitchell, *British Historical Statistics,* 258; Arthur Silverthorne, *The Purchase of Gas and Water Works, with the Latest Statistics of Municipal Gas and Water Supply* (London: Crosby Lockwood, 1881), 60–66; *Builder,* 21 Sept. 1901, 246.

19. Thomas Newbigging and W. T. Fewtrell, eds., *King's Treatise on the Science and Practice of the Manufacture and Distribution of Coal Gas,* 3 vols. (London: William B. King, 1878–82), esp. 1:109–13; Anthony S. Wohl, *Endangered Lives: Public Health in Victorian Britain* (Cambridge: Harvard University Press, 1983), 214.

20. Cotterill, "Development of Scottish Gas Technology," 23; Will Thorne, *My Life's Battles* (London: George Newnes, [1925]), 35–37, 51; Giles Radice and

Lisanne Radice, *Will Thorne, Constructive Militant: A Study in New Unionism and New Politics* (London: George Allen and Unwin, 1974).

21. George Livesey, *Report on the First 6 Months' Working of the Accident Fund* (South Metropolitan Gas Company, [1898]), 1–2, 9–11 (copy at the National Gas Archive, Warrington); Accident Report Book, Neepsend Works, Sheffield United Gas Company, 1887–1903, [EM] SHD/P/L/2, National Gas Archive, Warrington.

22. Letter to Walter Citrine, 31 May 1930, Modern Records Centre, University of Warwick, MS 292/144.5/6. See Thomas Oliver, "Coke-Men and By-Products Workers: Their Complaints and Maladies," *British Medical Journal*, 31 May 1930, 992–94; William A. Burgess, *Recognition of Health Hazards in Industry: A Review of Materials and Processes* (New York: John Wiley and Sons, 1981), 196; Barty-King, *New Flame*, 179, 191. Environmental historians have only relatively recently started tracing the effects of pollution back inside the factories from which they emanate. Two excellent examples of this approach are Christopher Sellers, "Factory as Environment: Industrial Hygiene, Professional Collaboration and the Modern Sciences of Pollution," *Environmental History Review* 18 (Spring 1994): 55–83; and Arthur F. McEvoy, "Working Environments: An Ecological Approach to Industrial Health and Safety," *Technology and Culture* 36, no. 2, supp. (1995): S145–72.

23. Newbigging and Fewtrell, *King's Treatise*, 1:119–20.

24. "No Headaches in Gas Works," *Gas Bulletin* 22 (1933): 57; BG11/CRG/A/A/78 and BG11/SE/ES/CRG/A/X/23, National Gas Archive, Warrington.

25. Thomas Bartlett Simpson, *Gas-Works: The Evils Inseparable from Their Existence in Populous Places. . . .* (London: William Freeman, 1866), 47; "The Gas," *Illustrated Times*, 5 Mar. 1864, 151, quoted in Lynda Nead, *Victorian Babylon: People, Streets, and Images in Nineteenth-Century London* (New Haven: Yale University Press, 2000), 94; "Smoke Prevention at Gas Works (A New Process of Gas Making)," *Lancet*, 12 Mar. 1904, 746.

26. William Hosgood Young Webber, *Town Gas and Its Uses for the Production of Light, Heat, and Motive Power* (London: Constable, 1907), 56, 94; Daniel Ellis, *Considerations Relative to Nuisance in Coal-Gas Works. . . .* (Edinburgh: John Anderson Jr., 1828), 18–19; Wilson, *Lighting the Town*, 35, 48; Earle B. Phelps, "Stream Pollution by Industrial Wastes and Its Control," in *A Half Century of Public Health*, ed. Mazÿck P. Ravenel, 197–208 (New York: American Public Health Association, 1921), 198; A. O. Thomas and J. N. Lester, "The Reclamation of Disused Gasworks Sites: New Solutions to an Old Problem," *Science of the Total Environment* 152, no. 3 (1994): 240–41. In the United States, by-products such as tar and ammonia remained unwanted until

the late nineteenth century. See Joel A. Tarr, "Transforming an Energy System: The Evolution of the Manufactured Gas Industry and the Transition to Natural Gas in the United States (1807–1954)," in *The Governance of Large Technical Systems*, ed. Olivier Coutard, 19–37 (London: Routledge, 1999), esp. 21. For an interesting example of the problems associated with reusing wastes in another context, see Timothy LeCain, "The Limits of 'Eco-Efficiency': Arsenic Pollution and the Cottrell Electrical Precipitator in the U.S. Copper Smelting Industry," *Environmental History* 5, no. 3 (2000): 336–51.

27. Quoted courtesy of Corporation of London, London Metropolitan Archives (B/NTG/2062 and B/NTG/2066); *Journal of Gas Lighting, Water Supply, &c.*, 4 July 1905, 52.

28. *Illustrated London News,* 11 Nov. 1865, quoted in Sarah Milan, "Refracting the Gaselier: Understanding Victorian Responses to Domestic Gas Lighting," in *Domestic Space: Reading the Nineteenth-Century Interior*, eds. Inga Bryden and Janet Floyd, 84–102 (Manchester: Manchester University Press, 1999), 98; *Times* (London), 10 Nov. 1865.

29. Williams, *A History of the British Gas Industry*, 26, 70; *Times* (London), 23 Mar. 1904; *London Argus*, 30 May 1902; *Kentish Mercury*, 31 Jan. 1902.

30. Walter Hepworth Collins, "The Alleged Danger to Public Health, Arising from Effluvium Nuisance from Gas Works," *Transactions of the Sanitary Institute of Great Britain* 11 (1890): 112–19, quotation from 113; Robert Roberts, *The Classic Slum: Salford Life in the First Quarter of the Century* (Manchester: Manchester University Press, 1971), 4; idem, *A Ragged Schooling: Growing Up in the Classic Slum* (Manchester: Mandolin, 1984), 133.

31. *Garston & Woolton Reporter,* 13 Jan. 1900, 2 Mar. 1901.

32. Ibid., 2 Mar. 1901.

33. Webber, *Town Gas,* 40–1; Birmingham City Council, Health Committee, Minutes, 10 Dec. 1894 (Birmingham City Archives); Williams, *A History of the British Gas Industry*, 62–63; United Kingdom Department of the Environment, *Problems Arising from the Redevelopment of Gas Works and Similar Sites*, 2nd ed. (London: HMSO, 1988), 22; "The Gas Council Memorandum on Air Pollution," typescript, n.d., Gas Council Records, Roll 152, National Gas Archive, Warrington.

34. Newbigging and Fewtrell, *King's Treatise,* 3:221.

35. William Henry Preece, "Gas versus Electricity," *Nature,* 23 Jan. 1879, 261–62; Leslie Hannah, *Electricity before Nationalisation: A Study of the Development of the Electricity Supply Industry in Britain to 1948* (Baltimore: Johns Hopkins University Press, 1979), 4; Charles William Siemens, *Science and Industry* (Birmingham, [1881]), 18.

36. "Exhibition of Gas Apparatus in Leeds," *Builder*, 10 May 1879, 510. Peak sales in Leeds exceeded 6 million cubic feet per day in the winter and were only 1.5 million cubic feet per day in the summer.

37. Philip Chantler, *The British Gas Industry: An Economic Study* (Manchester: Manchester University Press, 1938), 9; Hannah, *Electricity before Nationalisation*, 9; Wolfgang Schivelbusch, *Disenchanted Night: The Industrialization of Light in the Nineteenth Century*, trans. Angela Davies (Berkeley: University of California Press, 1988).

38. George Livesey, "Domestic Smoke Abatement," in *Addresses, Papers, and Discussions at the Conference on Smoke Abatement, Held in London, December 12–15th, 1905* (London: Royal Sanitary Institute, 1906), 13–19; Williams, *A History of the British Gas Industry*, 44.

39. Gwilym Gibbon and Reginald W. Bell, *History of the London County Council, 1889–1939* (London: Macmillan, 1939), 558.

40. Siemens, *Science and Industry*, 18.

41. *Times* (London), 21 Nov. 1901.

42. Roy Church, with the assistance of Alan Hall and John Kanefsky, *The History of the British Coal Industry*, vol. 3, *1830–1913: Victorian Pre-Eminence* (Oxford: Clarendon Press, 1986), 19; "Smoke Abatement," *Builder*, 10 Nov. 1883, 635; George Wyld, *Notes of My Life* (London: Kegan, 1903), 106.

43. "The Smoke-Abatement Exhibition," *Lancet*, 24 Dec. 1881, 1090.

44. Douglas Strutt Galton, *Fog and Smoke* (London, 1887), 17.

45. Lancet Analytical Sanitary Commission, "Second Report on Perfect Combustion and Smoke Prevention," *Lancet*, 5 Mar. 1892, 548–50, quotation on 548.

46. "The Smoke Abatement Exhibition," *Nature*, 5 Jan. 1882, 219.

47. Robert Lee to editor, *Lancet*, 30 Apr. 1892, 999.

48. Ellis Peyton to editor, *Lancet*, 10 Nov. 1906, 1305 (emphasis in original).

49. Thomas Coglan Horsfall, *The Nuisance of Smoke from Domestic Fires, and Methods of Abating It* (Manchester: Manchester and Salford Noxious Vapours Abatement Association, [1893?]), 7.

50. Williams, *A History of the British Gas Industry*, 10, 27; Robert Millward, "The Political Economy of Urban Utilities," in *The Cambridge Urban History of Britain*, vol. 3, *1840–1950*, ed. Martin Daunton, 315–49 (Cambridge: Cambridge University Press, 2000); Wilson, *Lighting the Town*, 194.

51. Ashby and Anderson, *Politics of Clean Air*, 42–43, 47; "Proceedings of the Health Section," *Sanitary Record*, 20 Oct. 1879, 140–41; Millward, "Political Economy," esp. 328.

52. "Manchester: Smoke," *Lancet*, 28 Apr. 1894, 1099–1100; Smoke Abatement League of Great Britain (Manchester and Salford Branch), *Case against*

the Levying of Contributions. . . . (Manchester, 1912), quoted in Mosley, *Chimney*, 158.

53. *Daily Telegraph* (London), 27 Apr. 1998.

54. "Gas Companies and Smoke Abatement," *Builder,* 6 Oct. 1883, 470; *Times* (London), 14 Dec. 1905. On the British Commercial Gas Association, see Chantler, *British Gas Industry,* 32–33; Williams, *A History of the British Gas Industry,* 74, 133.

55. Mitchell, *British Historical Statistics,* 269; Compton Mackenzie, *The Vital Flame* (London: British Gas Council, 1947), 21–23; *Times* (London), 24 Nov. 1898. Yet most people continued to use solid fuels, especially bituminous coal, for heating. When the gas industry was nationalized, gas was providing less than 15 percent of the heat used in British houses. See Barty-King, *New Flame,* 232.

56. Webber, *Town Gas,* 227; "Smoke Abatement," *Gas Bulletin* 11 (July 1922): 141; *Times* (London), 18 Sept. 1923; Francis William Goodenough, "The Fuel of the Future," in Smoke Abatement League of Great Britain, *Report of the Smoke Abatement Conference held at the Town Hall, Manchester, November 4th, 5th and 6th, 1924,* 248–54 (Manchester, 1924), 254.

57. National Smoke Abatement Society, Minutes, 5 Oct. 1929, National Society for Clean Air and Environmental Protection, Brighton; *Times* (London), 7 Oct. 1929, 11b.

58. J. S. Taylor, *Smoke and Health* (Manchester: National Smoke Abatement Society, 1929), 3.

59. National Smoke Abatement Society, Minutes, 24 Apr. 1930, National Society for Clean Air and Environmental Protection, Brighton.

60. *Times* (London), 22 Sept. 1936, 7f.

61. National Smoke Abatement Society, *Smoke Abatement Exhibition Handbook and Guide* (Manchester: National Smoke Abatement Society, 1936), 60; Z Archive, Z111/8, negs. 8604, 8609, Science Museum Library, Archives Collection, London; "Proposed New Sections," typescript, June 1937, ibid., Z186; Williams, *A History of the British Gas Industry,* 74; "Smoke into Sales," *Gas Bulletin* 28 (Feb. 1939): 32; Timothy Boon, "'The Smoke Menace': Cinema, Sponsorship and the Social Relations of Science in 1937," in *Science and Nature: Essays in the History of the Environmental Sciences,* ed. Michael Shortland, 57–88 ([Stanford in the Vale]: British Society for the History of Science, 1993), esp. 85.

62. National Smoke Abatement Society, Minutes, 16 Sept. 1938, National Society for Clean Air and Environmental Protection, Brighton; Arnold Marsh, *Smoke: The Problem of Coal and the Atmosphere* (London: Faber and Faber, [1947]), facing 260–61.

63. Department of Scientific and Industrial Research, *The Investigation of Atmospheric Pollution: A Report on Observations in the 5 Years Ended 31st March, 1944* (London: HMSO, 1949), 69; Henry T. Bernstein, "The Mysterious Disappearance of Edwardian London Fog," *London Journal* 1, no. 2 (1975): 189–206.

64. *Guardian* (London), 21 Mar. 1998, 15 Nov. 2001; *Sunday Times* (London), 3 June 2001; Paul Goldberger, "The Big Top," *New Yorker,* 27 Apr./4 May 1998, 152–59.

65. William Byers, Martin B. Meyers, and Donna E. Mooney, "Analysis of Soil from a Disused Gasworks," *Water, Air, and Soil Pollution* 73 (1994): 1–9; Thomas and Lester, "Reclamation of Disused Gasworks Sites"; United Kingdom Department of the Environment, *Problems Arising,* esp. 35; Sean Humber, *Gas Works Sites in London: An Investigation into Contaminated Land* (London: Friends of the Earth, 1991), esp. 24–25; Neil S. Shifrin et al., "Chemistry, Toxicity, and Human Health Risk of Cyanide Compounds in Soils at Former Manufactured Gas Plant Sites," *Regulatory Toxicology and Pharmacology* 23, no. 2 (1996): 106–16.

Chapter 10

1. Quoted in *Daily Mirror* (London), 8 Dec. 1952.

2. An earlier version of this chapter was published as "Interpreting the London Fog Disaster of 1952," in *Smoke and Mirrors: The Politics and Culture of Air Pollution,* ed. E. Melanie DuPuis, 154–69 (New York: New York University Press, 2004).

3. National Smoke Abatement Society, *Proceedings of the London Conference, 1943,* 8 (copy at National Society for Clean Air and Environmental Protection, Brighton).

4. National Smoke Abatement Society, *Smoke Abatement in Wartime* (1940), 3–4 (copy at National Society for Clean Air and Environmental Protection, Brighton).

5. W. A. Damon, "Smoke Abatement," 3 Apr. 1942, MH 58/398, National Archives.

6. Clough Williams-Ellis, "Unholy Smoke," in National Smoke Abatement Society, *Proceedings of the London Conference, 1943,* 3 (emphasis in original).

7. National Smoke Abatement Society, *Smoke Prevention in Relation to Initial Post-War Reconstruction* (1942), 2.

8. *Daily Telegraph* (London), 2 Oct. 1953; *Times* (London), 25 Nov. 1952, 5; "Big Words, Tiny Coal," *Economist,* 26 Oct. 1957, 335. On the rationing of coal, see J. C. R. Dow, *The Management of the British Economy, 1945–60* (Cam-

bridge: Cambridge University Press, 1964), 148. On the confrontation between the National Coal Board and the NSAS, see National Smoke Abatement Society, Technical Committee, Minutes, 19 Dec. 1952, National Society for Clean Air and Environmental Protection, Brighton.

9. *Daily Mirror* (London), 8 Dec. 1952.

10. C. K. M. Douglas and K. H. Stewart, "London Fog of December 5–8, 1952," *Meteorological Magazine* 82 (1953): 67–71; E. T. Wilkins, "Air Pollution and the London Fog of December, 1952," *Journal of the Royal Sanitary Institute* 74 (Jan. 1954): 1–21; Ministry of Health, *Mortality and Morbidity during the London Fog of December 1952* (London: HMSO, 1954), table 3.

11. *Times* (London), 6 Dec. 1952; ibid., 8 Dec. 1952; ibid., 9 Dec. 1952.

12. G. F. Abercrombie, "December Fog in London and the Emergency Bed Service," *Lancet,* 31 Jan. 1953, 234–35.

13. John Fry, "Effects of a Severe Fog on a General Practice," *Lancet,* 31 Jan. 1953, 235–36.

14. House of Commons, *Parliamentary Debates,* 16 Dec. 1952, col. 188; ibid., 17 Dec. 1952, col. 221; ibid., 18 Dec. 1952, col. 237; *Daily Telegraph* (London), 19 Dec. 1952; "Deaths in the Fog," *British Medical Journal,* 3 Jan. 1953, 50.

15. House of Commons, *Parliamentary Debates,* 22 Jan. 1953, col. 382.

16. *Evening Standard* (London), 24 Jan. 1953, quoted in House of Commons, *Parliamentary Debates,* 8 May 1953, cols. 842–43.

17. House of Commons, *Parliamentary Debates,* 27 Jan. 1953, cols. 828, 830; ibid., 12 Feb. 1953, col. 75; ibid., 19 Mar. 1953, col. 189.

18. Ibid., 8 May 1953, cols. 841, 846, 850.

19. Committee on Air Pollution, *Report,* Cmd. 9322 (1954); *Times* (London), 18 Jan. 1967.

20. Harold Macmillan, letter to David Maxwell Fyfe, 17 Nov. 1953, MH 58/398, National Archives.

21. Ministry of Health, *Mortality and Morbidity,* table 1.

22. *Times* (London), 31 Jan. 1953; "The Toll of Fog," *British Medical Journal,* 7 Feb. 1953, 321.

23. Ministry of Health, *Mortality and Morbidity,* 2.

24. W. P. D. Logan, "Mortality in the London Fog Incident, 1952," *Lancet,* 14 Feb. 1953, 336–38, esp. 337.

25. House of Commons, *Parliamentary Debates,* 21 Jan. 1953, col. 42.

26. Logan, "Mortality," 336 (emphasis added).

27. Wilkins, "Air Pollution," 11.

28. Ministry of Health, *Mortality and Morbidity,* 14; Michelle L. Bell and Devra Lee Davis, "Reassessment of the Lethal London Fog of 1952: Novel

Indicators of Acute and Chronic Consequences of Acute Exposure to Air Pollution," *Environmental Health Perspectives* 109, supp. 3 (June 2001): 389–94. See also Devra Lee Davis, *When Smoke Ran Like Water: Tales of Environmental Deception and the Battle against Pollution* (New York: Basic Books, 2002), 42–54.

29. House of Commons, *Parliamentary Debates,* 29 Oct. 1953, col. 407.

30. Quoted in *Times* (London), 31 Jan. 1953.

31. "The Toll of Fog," 321.

32. "The Menace of Air Pollution," *Planning,* 16 Aug. 1954, 189–216, quotation on 193–94. Similar debates about the "reality" of deaths attributed to public health disasters have occurred in other contexts. See Eric Klinenberg, *Heat Wave: A Social Autopsy of Disaster in Chicago* (Chicago: University of Chicago Press, 2002), esp. 24–31.

33. House of Commons, *Parliamentary Debates,* 13 Nov. 1953, col. 105.

34. *Times* (London), 20 Apr. 1953.

35. Committee on Air Pollution, *Report,* Cmd. 9322 (1954), 8. The committee's attention to the high rate of bronchitis in Britain relative to other countries was likely influenced by Neville C. Oswald, James T. Harold, and W. J. Martin, "Clinical Pattern of Chronic Bronchitis," *Lancet,* 26 Sept. 1953, 639–43.

36. Committee on Air Pollution, *Report,* Cmd. 9322 (1954), 8.

37. United Nations, *Demographic Yearbook, 1954* (New York: United Nations, 1954).

38. Committee on Air Pollution, *Report,* Cmd. 9322 (1954), 9.

39. Hugh Beaver, typescript of a speech to the Engineering Institute of Canada, 17 Mar. 1955, Beaver Papers, British Library of Political and Economic Science, London.

40. Roy M. Harrison, "Important Air Pollutants and Their Chemical Analysis," in *Pollution: Causes, Effects, and Control,* 2nd ed., ed. R. M. Harrison (Cambridge: Royal Society of Chemistry, 1990), 139.

Chapter 11

1. Harvey Littlejohn, *Report on the Causes and Prevention of Smoke from Manufacturing Chimneys* (1897), 21 (copy at Sheffield Local Studies Library).

2. H. A. Des Voeux, *Smoke Abatement* (Manchester: National Smoke Abatement Society, 1936), 6.

3. Committee on Air Pollution, *Report,* Cmd. 9322 (1954), 5.

4. *Times* (London), 20 Apr. 1953.

5. Hugh Beaver, typescript of a speech delivered at the University of London, 1 Dec. 1955, 4, Beaver Papers, British Library of Political and Economic Science, London (hereafter Beaver Papers).

6. E. T. Wilkins, "Air Pollution and the London Fog of December, 1952," *Journal of the Royal Sanitary Institute* 74 (Jan. 1954): 14.

7. Hugh Beaver, typescript copy of a speech delivered in New York, 2 Mar. 1955, 11–12, Beaver Papers.

8. Committee on Air Pollution, *Interim Report,* Cmd. 9011 (1953); "Committee on Air Pollution: Notes" [1954], POWE 28/246, National Archives.

9. Committee on Air Pollution, *Report,* Cmd. 9322 (1954), 44–45.

10. Ibid., 21.

11. *Times* (London), 27 Oct. 1936, 9c; National Smoke Abatement Society, *Smoke Control Areas* (London: NSAS, [ca. 1956]).

12. Hugh Beaver, typescript copy of a speech to the Engineering Institute of Canada, 17 Mar. 1955, Beaver Papers.

13. G. E. Foxwell, "Smoke Abatement in Practice," *Journal of the Institute of Fuel* 28 (Sept. 1955): 451–57, quotation from 452.

14. Beaver, Engineering Institute of Canada, 17 Mar. 1955.

15. Foxwell, "Smoke Abatement in Practice," 452.

16. "Committee on Air Pollution: Notes," HLG 55/78, National Archives.

17. Committee on Air Pollution, *Report,* Cmd. 9322 (1954), 7.

18. North Thames Gas Board, letter to Gas Council, 2 Dec. 1952; "The Gas Council Memorandum on Air Pollution," typescript, n.d. (both in Gas Council Records, Roll 152, National Gas Archive, Warrington).

19. A. E. Haffner, "The Control of Atmospheric Pollution by Gasworks," address to the London and Southern Section of the Institution of Gas Engineers, 21 Apr. 1959 (copy at National Gas Archive, Warrington).

20. F. Reynolds, *The Law Relating to Air Pollution* (Brighton: National Society for Clean Air, n.d.), 21; William Ashworth and Mark Pegg, *The History of the British Coal Industry,* vol. 5, *1946–1982: The Nationalized Industry* (Oxford: Clarendon Press, 1986), 504.

21. Ministry of Housing and Local Government, "Beaver Report," 27 Jan. 1955, POWE 28/257, National Archives.

22. Philip Chantler, "Beaver Report," 20 Jan. 1955, POWE 28/257, National Archives.

23. Hugh Beaver, copy of letter to Rose Cave, 17 Aug. 1955, Beaver Papers.

24. Committee on Air Pollution, *Report,* Cmd. 9322 (1954), 17–18.

25. Ibid., 17; "Home Affairs Committee: Report," ca. Dec. 1954, HLG 55/78, National Archives (emphasis in original).

26. Committee on Air Pollution, *Report,* Cmd. 9322 (1954), 18.

27. North Thames Gas Board, Fulham Laboratory, "Sulphur Balance for Gas Works," 30 Nov. 1953, Gas Council Records, Roll 152, National Gas Archive.

28. Committee on Air Pollution, *Report,* Cmd. 9322 (1954), 6.

29. "Air Pollution," *British Medical Journal,* 4 Dec. 1954, 1341.

30. "Discussion on 'The Beaver Report,'" *Journal of the Institute of Fuel* 28 (Apr. 1955): 193.

31. Ibid., 188, 194.

32. "Home Affairs Committee: Report," HLG 55/78, National Archives.

33. Hugh Beaver, typescript extract of letter to Duncan Sandys, 5 June 1955, Beaver Papers; Duncan Sandys, letter to Hugh Beaver, 4 July 1955, Beaver Papers; National Smoke Abatement Society, *Smoke Control Areas.*

34. Hugh Beaver to Duncan Sandys, 5 June 1955, Beaver Papers.

35. "Coal's Case to Answer," *Economist,* 13 Feb. 1965, 640–41.

36. "The Chamber of Coal Traders: Note of Suggestions," 25 Oct. 1955, POWE 28/257, National Archives.

37. Gas Council, meeting notes, typescript, 14 Jan. 1954, Gas Council Records, Roll 152, National Gas Archive, Warrington.

38. National Smoke Abatement Society, *Smoke Control Areas.*

39. Duncan Sandys to Hugh Beaver, 4 July 1955, Beaver Papers.

40. National Smoke Abatement Society, "Minutes of a Meeting of a Sub-Committee of the Technical Committee," 21 Sept. 1955, National Society for Clean Air and Environmental Protection, Brighton.

41. Michael Graham, *Wage War on Smog* (London: Conservative Political Centre, 1959), 17.

42. Salford Health Department, *Smoke Control Means Clean Air for You and Your Family,* n.d. (copy at National Society for Clean Air and Environmental Protection, Brighton).

43. National Smoke Abatement Society, *Smoke Control Areas.*

44. Ministry of Housing and Local Government, *Smoke Control (England & Wales): Summary of Programmes Submitted by Local Authorities for the Establishment of Smoke Control Areas,* Cmnd. 1113 (1960), 3–5.

45. James H. Winter, *Secure from Rash Assault: Sustaining the Victorian Environment* (Berkeley: University of California Press, 1999), 143–46; John Sheail, *An Environmental History of Twentieth-Century Britain* (New York: Palgrave, 2002), 55–56, 231–35.

46. Roy Church, with the assistance of Alan Hall and John Kanefsky, *The History of the British Coal Industry,* vol. 3, *1830–1913: Victorian Pre-Eminence* (Oxford: Clarendon Press, 1986), 85, 91–92.

47. Barry Supple, *The History of the British Coal Industry,* vol. 4, *1913–1946: The Political Economy of Decline* (Oxford: Clarendon Press, 1987), 15.

48. *Times* (London), 14 Sept. 1954, 4c.

49. Leonora Murray, "The Miner's Home," *Nineteenth Century and After,* Jan. 1929, 9–31, quotations on 9–11.

50. Norman Dennis, Fernando Henriques, and Clifford Slaughter, *Coal Is Our Life: An Analysis of a Yorkshire Mining Community* (London: Eyre and Spottiswoode, 1956).

51. James Law, "Atmospheric Pollution," *Journal of the Royal Sanitary Institute* 5 (Nov. 1939): 202–13, quotation on 203.

52. B. W. Clapp, *An Environmental History of Britain since the Industrial Revolution* (New York: Longman, 1994), 162.

53. J. E. C. Munro, "Return of Rates of Wages in the Mines and Quarries in the United Kingdom with Report thereon of 1891 (C-6455)," *Economic Journal* 1, no. 4 (1891): 806–8. On the utilization of previously unsalable materials in the meatpacking industry, see William Cronon, *Nature's Metropolis: Chicago and the Great West* (New York: W. W. Norton, 1991), 250–53.

54. Similar arrangements existed in the U.S. lumber industry during this period. See Cronon, *Nature's Metropolis,* 164–67.

55. *Times* (London), 11 May 1949, 5c.

56. Ashworth and Pegg, *History of the British Coal Industry,* 5: 672.

57. *Times* (London), 17 Feb. 1949, 4f.

58. Ministry of Housing and Local Government, *Smoke Control (England & Wales),* 3–5; "Committee on Air Pollution: Notes by the Chairman of the Gas Council" [1954], Gas Council Records, Roll 152, National Gas Archive; Ashworth and Pegg, *History of the British Coal Industry,* 210.

59. *Times* (London), 1 June 1949, 2c; Daniel Davies, "Coal and Its Producers in Wartime," *Quarterly Review,* July 1944, 284–97, esp. 296.

60. Ashworth and Pegg, *History of the British Coal Industry,* 5: 103.

61. *Times* (London), 23 Aug. 1956, 6c; ibid., 3 Sept. 1956, 4c. See also ibid., 10 Oct. 1949, 4e.

62. Church, Hall, and Kanefsky, *History of the British Coal Industry,* 3: 93.

63. *Times* (London), 12 Nov. 1959, 5e.

64. National Smoke Abatement Society, Technical Committee, Minutes, 21 Jan. 1957, National Society for Clean Air and Environmental Protection, Brighton.

65. "Home Fires Burning Bright," *Economist,* 27 May 1961, 901.

66. Ministry of Power, *Domestic Fuel Supplies and the Clean Air Policy,* Cmnd. 2231 (1963), 8.

67. S. R. Craxford and Marie-Louise P. M. Weatherley, "Air Pollution in Towns in the United Kingdom," *Philosophical Transactions of the Royal Society of London. Series A, Mathematical and Physical Sciences* 269, no. 1199 (1971): 503–13, esp. 503, 507, 508; Howard A. Scarrow, "The Impact of British Domestic Air Pollution Legislation," *British Journal of Political Science* 2, no. 3 (1972): 261–82, esp. 279.

68. Craxford and Weatherley, "Air Pollution in Towns," 507; Scarrow, "Air Pollution Legislation," 279.

69. *Times* (London), 18 Oct. 1967, 4e.

70. "Home Fires," *Economist,* 901–2.

71. Craxford and Weatherley, "Air Pollution in Towns," 505; B. R. Mitchell, *British Historical Statistics* (Cambridge: Cambridge University Press, 1988), 259.

72. Craxford and Weatherley, "Air Pollution in Towns," 506; Scarrow, "Air Pollution Legislation," 261–82, esp. 271 n 23.

73. Albert Parker, "What's in the Air?" *Memoirs and Proceedings of the Manchester Literary and Philosophical Society* 111 (1968–69): 48–63, esp. 51–53; Mitchell, *British Historical Statistics,* 259; Ashworth and Pegg, *History of the British Coal Industry,* 5: 678–79.

74. Brian Joseph McCormick, *Industrial Relations in the Coal Industry* (London: Macmillan, 1979), 229.

75. *Times* (London), 28 Jan. 1975, 4b.

76. Ashworth and Pegg, *History of the British Coal Industry,* 5: 673, 678; Joe Hicks and Grahame Allen, *A Century of Change: Trends in UK Statistics since 1900* (London: House of Commons Library, 1999).

77. *Times* (London), 28 Jan. 1975, 4b; Department of Trade and Industry, "Energy Minister Acts to Protect Supplies of Concessionary Fuel," press release, 16 Dec. 1999, accessed online via Lexis/Nexis; *Journal* (Newcastle), 11 Oct. 1999, 3.

78. *Scottish Daily Record,* 30 Aug. 1995, 15.

79. Eric Ashby and Mary Anderson, *The Politics of Clean Air* (Oxford: Clarendon Press, 1981).

80. W. Idris Jones, "Fuel and Power Research in Britain," *Journal of the Institute of Fuel* 28 (1955): 262–71, esp. 269.

81. Michael Stratton and Barrie Stuart Trinder, *Twentieth Century Industrial Archaeology* (London: E. & F. N. Spon, 2000), 31–32.

82. "Home Affairs Committee: Report," HLG 55/78, National Archives.

83. G. E. Foxwell, "Smoke Abatement in Practice," *Journal of the Institute of Fuel* 28 (Sept. 1955): 451–57, quotation from 456.

84. M. J. Edwards, *Clean Air and Solid Fuel,* (London: National Coal Board, 1970), 3 (copy at National Society for Clean Air and Environmental Protection, Brighton); Ashworth and Pegg, *History of the British Coal Industry,* 5: 495.

85. National Society for Clean Air, *Notes on the Background to Clean Air. . . .* (Brighton: National Society for Clean Air, [ca. 1970]), 9.

Conclusion

1. Mary Douglas, *Purity and Danger: An Analysis of Concepts of Pollution and Taboo* (London: Routledge and Kegan Paul, 1966).

2. See the classic essay by Raymond Williams, "Ideas of Nature," in *Problems in Materialism and Culture* (London: Verso, 1980), 67–85.

3. B. R. Mitchell, *International Historical Statistics: Europe, 1750–1993* (London: Macmillan Reference, 1998), 485.

4. Chris Cook and John Stevenson, *The Longman Handbook of Modern British History, 1714–2001,* 4th ed. (London: Longman, 2001), 257.

5. Department of Trade and Industry, *UK Energy in Brief* (2004), 17–18.

6. John Sheail, *An Environmental History of Twentieth-Century Britain* (New York: Palgrave, 2002), 254–55.

7. J. R. McNeill, *Something New under the Sun: An Environmental History of the Twentieth-Century World* (New York: W. W. Norton, 2000), 108–11.

8. Intergovernmental Panel on Climate Change, *Climate Change: The IPCC Scientific Assessment,* ed. J. T. Houghton, G. J. Jenkins, and J. J. Ephraums (New York: Cambridge University Press, 1990).

Bibliography

Primary Sources

Archives Consulted

Birmingham City Archives
Bodleian Library, Special Collections and Western Manuscripts, Oxford
    University
British Engine Insurance, Ltd.
British Library of Political and Economic Science, London School of
    Economics
British Library, Department of Manuscripts
British Medical Association
City of Westminster Archives Centre
Glasgow City Archives
Glasgow University Archive Services
Guildhall Library
Hammersmith and Fulham Archives and Local History Centre
Institution of Civil Engineers
John Rylands University Library, University of Manchester
King Alfred School, London
London Metropolitan Archives
Manchester Archives and Local Studies
Modern Records Centre, University of Warwick
Museum of English Rural Life, University of Reading
National Archives of England, Wales and the United Kingdom
National Archives of Ireland
National Archives of Scotland
National Gas Archive

National Library of Scotland, Department of Manuscripts
National Society for Clean Air and Environmental Protection
National Trust
Royal Academy of Arts
Royal College of Physicians and Surgeons of Glasgow
Royal College of Physicians of London
Royal Society for the Promotion of Health
Science Museum Library, Archives Collection
Sheffield Archives
Surrey History Centre
University College London, Manuscripts Room
Wellcome Library for the History and Understanding of Medicine
West Yorkshire Archive Service, Leeds
Working Class Movement Library

Parliamentary Papers and Government Reports

Committee on Air Pollution. *Report.* Cmd. 9322. H.C. 1954.
Department of Scientific and Industrial Research. *The Investigation of Atmospheric Pollution: A Report on Observations in the 5 Years Ended 31st March, 1944.* London: HMSO, 1949.
Department of the Environment. *Problems Arising from the Redevelopment of Gas Works and Similar Sites,* 2nd ed. London: HMSO, 1988.
Department of Trade and Industry. "Energy Minister Acts to Protect Supplies of Concessionary Fuel." Press release, 16 Dec. 1999. Accessed online via LexisNexis.
Department of Trade and Industry. *UK Energy in Brief.* 2004.
House of Commons. *Parliamentary Debates.*
Inter-Departmental Committee on Physical Deterioration. *Report [and Minutes of Evidence].* Cd. 2175, Cd. 2210. H.C. 1904.
Ministry of Health. *Interim Report of the Committee on Smoke and Noxious Vapours Abatement.* Cmd. 755. H.C. 1920.
Ministry of Health. *Mortality and Morbidity during the London Fog of December 1952.* London: HMSO, 1954.
Ministry of Housing and Local Government. *Smoke Control (England and Wales): Summary of Programmes Submitted by Local Authorities for the Establishment of Smoke Control Areas.* Cmnd. 1113. H.C. 1960.
Ministry of Power. *Domestic Fuel Supplies and the Clean Air Policy,* Cmnd. 2231. London: HMSO, 1963.
Report from the Select Committee of the House of Lords on Smoke Nuisance Abatement (Metropolis) Bill [H.L.]. H.C. 1887.

Report from the Select Committee on Coal. H.C. 1873.

Report of the Commissioners Appointed to Inquire into the Several Matters Relating to Coal in the United Kingdom, iii. C. 435-2. H.C. 1871.

Report of the Select Committee of the House of Lords on Injury from Noxious Vapours. H.L. 1862.

Reports on the Laws in Force in Certain Foreign Counties in Regard to the Emission of Smoke from Chimneys. Cd. 2347. H.C. 1905.

Smoke Control (England & Wales): Summary of Programmes Submitted by Local Authorities for the Establishment of Smoke Control Areas. Cmnd. 1113. H.C. 1960.

Newspapers Consulted

Daily Mail (London)
Daily News (London)
Daily Telegraph (London)
Evening Standard (London)
Garston & Woolton Reporter
Glasgow Herald
Illustrated London News
Kentish Mercury
London Argus
Manchester Guardian
Sheffield Daily Telegraph
Times (London)

Books and Articles

Abercrombie, G. F. "December Fog in London and the Emergency Bed Service." *Lancet,* 31 Jan. 1953, 234–35.

Addy, S. O., and W. T. Pike. *Sheffield at the Opening of the Twentieth Century: Contemporary Biographies.* Brighton: W. T. Pike, n.d.

Aikman, C. M. *Air, Water, and Disinfectants.* Manuals of Health. London: Society for Promoting Christian Knowledge, 1895.

"The Air of Glasgow." *Sanitary Record,* 15 Jan. 1880, 261.

Alden, Percy. "Coal Smoke Abatement." *Contemporary Review* 122 (Dec. 1922): 725–33.

Alison, S. Scott. *An Inquiry into the Propagation of Contagious Poisons, by the Atmosphere; as also into the Nature and Effects of Vitiated Air, Its Forms and Sources, and Other Causes of Pestilence. . . .* Edinburgh: Maclachan, Stewart, 1839.

"The Analysis of the Air of Large Cities." *Industries,* 23 Jan. 1891, 91.

Angell, John. "Address on the Paris Electrical and London Smoke-Abatement Exhibitions." *Report and Proceedings of the Manchester Field-Naturalists and Archaeologists' Society for the Year 1882* (1882): 55–60.

———. *Personal and Household Arrangements in Relation to Health.* Manchester and Salford Sanitary Association, [ca. 1878].

A.R.I.B.A. "The Obelisk." *Builder,* 23 Feb. 1878, 198.

Arnold, R. Arthur. *Sanitary Reform: "Water Supply" and "Pure Air."* London: Bradbury, Evans, 1866.

Arnott, Neil. *On the Smokeless Fire-Place, Chimney-Valves, and Other Means, Old and New, of Obtaining Healthful Warmth and Ventilation.* London: Longmans, Brown, Green, and Longmans, 1855.

Arrhenius, Svante. "On the Influence of Carbonic Acid in the Air upon the Temperature of the Ground." *London, Edinburgh, and Dublin Philosophical Magazine,* 5th ser., 41 (Apr. 1896): 237–76.

Ascher, Louis. "Coal Smoke Abatement in England." *Journal of the Royal Sanitary Institute* 28 (Mar. 1907): 88–93.

Atkinson, John Charles. *Change of Air: Fallacies Regarding It.* London: John Ollivier, 1848.

Bailey, William H. "Open Spaces and Recreational Grounds." *Report and Proceedings of the Manchester Field-Naturalists and Archaeologists' Society for the Year 1883* (1884): 56–66.

Barker, Thomas Herbert. *On Malaria and Miasmata, and Their Influence in the Production of Typhus and Typhoid Fevers, Cholera, and the Exanthemata.* London: John W. Davies, 1863.

Barlow, Peter W. *Smoke Abatement.* London: Vacher, 1882.

Barnett, Henrietta O. *Canon Barnett: His Life, Work, and Friends,* 2 vols. London: John Murray, 1919.

Barr, Robert. "The Doom of London." *Idler* 2 (1892–93): 397–409. Reprinted in *The Smoake of London: Two Prophecies,* edited by James P. Lodge. Elmsford, N.Y.: Maxwell Reprint, 1969.

Barwise, Sidney. "The Abatement of the Smoke Nuisance." *Sanitary Record,* 15 June 1891, 609–13.

Bascome, Edward. *Prophylaxis, or the Mode of Preventing Disease by a Due Appreciation of the Grand Elements of Vitality: Light, Air, and Water, with Observations on Intramural Burials.* London: S. Highley, 1849.

Bass, Thomas J. *Tragedies of Life: A Fragment of To-day.* Birmingham, [ca. 1903].

Bell, J. Carter. *Noxious Vapours Which Pollute the Air.* Manchester: Manchester and Salford Noxious Vapours Abatement Association, [1888?].

"Big Words, Tiny Coal." *Economist*, 26 Oct. 1957, 335.

Birmingham Ladies' Association for Useful Work. *Annual Report*, 1880. Copy at Local Studies and History Service, Birmingham.

Blasius, William. "Some Remarks on the Connection of Meteorology with Health." *Proceedings of the American Philosophical Society* 14 (1875): 667–71.

Blyth, A. Wynter. "Ventilation." *Public Health* 14 (1901–2): 61–91.

Blyth, Lindsey. *Minute of Information on Disinfection and Deodorization*. London: HMSO, 1857.

Booth, William. *In Darkest England, and The Way Out*. London: Salvation Army, 1890. Reprinted Montclair, N.J.: Patterson Smith, 1975.

Bousfield, William. "Smoke in the Manufacturing Districts." *Art Journal*, 1882, 9–10.

Brabazon, Reginald, Twelfth Earl of Meath. "Decay of Bodily Strength in Towns." *Nineteenth Century* 21 (May 1887): 673–76.

———. "Great Cities and Social Reform, I." *Nineteenth Century* 14 (Nov. 1883): 798–809.

———. "Health and Physique of Our City Populations." *Nineteenth Century* 10 (July 1881): 80–89.

———. "The London County Council and Open Spaces." *New Review* 7 (Dec. 1892): 701–7.

———. "Lungs for Our Great Cities." *New Review* 2 (May 1890): 432–43.

———. *Physical Training in Relation to the Needs of a Town Population*. [1902].

———. *Social Arrows*. London: Longmans, Green, 1886.

Brodie, Frederick J. "Decrease of Fog in London during Recent Years." *Quarterly Journal of the Royal Meteorological Society* 31 (1905): 15–28.

———. "On the Prevalence of Fog in London during the Years 1871 to 1890." *Quarterly Journal of the Royal Meteorological Society* 18 (1892): 40–45.

Bruce, Eric Stuart. "Town Fogs: Their Amelioration and Prevention." *Dublin Review* 114 (Jan.–Apr. 1894): 132–44.

Carpenter, Alfred. "The First Principles of Sanitary Work." *Sanitary Record*, 15 Dec. 1879, 203–4.

———. "London Fogs." *Journal of the Society of Arts*, 10 Dec. 1880, 48–60.

———. *Preventive Medicine in Relation to the Public Health*. London: Simpkin, Marshall, 1877.

Carpenter, Edward. "Coal and Wet Weather: An Object Lesson from Sheffield." *Daily News* (Sheffield), 21 June 1921.

———. "The Smoke-Plague and Its Remedy." *Macmillan's Magazine* 62 (July 1890): 204–13.

———. "Sunshine and Coal." *Daily News* (Sheffield), 28 May 1921.

"Casualties Due to Fog and Frost." *Lancet*, 2 Jan. 1892, 33.

Chadwick, Edwin. *Report on the Sanitary Condition of the Labouring Population of Great Britain.* 1842. Edited by M. W. Flinn. Edinburgh: Edinburgh University Press, 1965.

Chalmers, A. K. *The Health of Glasgow, 1818–1925: An Outline.* Glasgow: Corporation [of the City], 1930.

Chambers, William. "A London Fog." *Chambers's Journal*, 4 Dec. 1880, 769–71.

Chaumont, [Francis Stephen Bennet] François de. *The Habitation in Relation to Health.* London: Society for Promoting Christian Knowledge, 1879.

The Cheap Doctor: A Word about Fresh Air. London: Ladies' National Association for the Diffusion of Sanitary Knowledge, [1859].

"Chemical and Physical Modifications of the Atmosphere Consequent on Habitation." *Chemical News*, 15 Mar. 1862, 146–48.

Chubb, Lawrence W. *Smoke Abatement.* London, 1912. Copy at National Society for Clean Air and Environmental Protection, Brighton.

"Cleopatra's Needle." *British Architect and Northern Engineer*, 23 Mar. 1877, 174–75.

"Clergymen as Sanitarians." *Sanitary Record*, 15 May 1881, 423–24.

Coal Smoke Abatement Society. *Summary of Law Relating to Smoke Pollution.* London, 1901. Copy at National Society for Clean Air and Environmental Protection, Brighton.

———. *Papers Read at the Smoke Abatement Conferences, March 26, 27, and 28, 1912, with Discussions.* London, 1912.

"The Coal Smoke Abatement Society and Domestic Grates." *Lancet*, 10 May 1902, 1332–35.

"Coal's Case to Answer." *Economist*, 13 Feb. 1965, 640–41.

Cobb, John W. "Coal Conservation." *Edinburgh Review* 229 (Jan. 1919): 39–61.

Cohen, Julius B. *The Air of Towns.* Washington: Government Printing Office, 1896.

———. *The Character and Extent of Air Pollution in Leeds.* Leeds, 1896.

———. "A Record of the Work of the Leeds Smoke Abatement Society." In *Addresses, Papers, and Discussions at the Conference on Smoke Abatement, Held in London, December 12–15th, 1905*, 27–29. London: Royal Sanitary Institute, 1906.

Cohen, Julius B., and Arthur G. Ruston. *Smoke: A Study of Town Air.* London: Edward Arnold, 1912.

Coles, W. R. E. "Smoke Abatement." *Transactions of the Sanitary Institute of Great Britain* 5 (1883): 335–45.

———. "Smoke Abatement." *Transactions of the Sanitary Institute of Great Britain* 7 (1885): 218–28.

Collins, John. "Air and Ventilation." *Transactions of the Sanitary Institute of Great Britain* 6 (1884–85): 391–95.

Collins, Walter Hepworth. "The Alleged Danger to Public Health, Arising from Effluvium Nuisance from Gas Works." *Transactions of the Sanitary Institute of Great Britain* 11 (1890): 112–19.

Colvin, Sidney. "Restoration and Anti-Restoration." *Nineteenth Century* 2 (Oct. 1877): 456–61.

Commons Preservation Society. *Report of Proceedings*, 1882–83.

"Condition of Our Chief Towns—Sheffield." *Builder*, 21 Sept. 1861, 641.

Condy, Henry Bollmann. *Air and Water: Their Impurities and Purification.* London: John W. Davies, 1862.

Cooke, Nicholas. "Melanism in Lepidoptera." *Entomologist* 10 (June 1877): 151–53.

———. "On Melanism in Lepidoptera." *Entomologist* 10 (Apr. 1877): 92–96.

"The Crime of Being Inefficient." *Nation* (London), 25 May 1912, 275–77; 15 June 1912, 390–92.

Crookes, William. *The Wheat Problem, Based on Remarks Made in the Presidential Address to the British Association in Bristol in 1898.* London: John Murray, 1899.

Daly, Martin H. "The Smoke Nuisance." *Brewers' Journal*, 15 Feb. 1909, 124–25.

Davies, Daniel. "Coal and Its Producers in Wartime." *Quarterly Review*, July 1944, 284–97.

"Deaths in the Fog." *British Medical Journal*, 3 Jan. 1953, 50.

Des Voeux, H. A. *Smoke Abatement.* Manchester: National Smoke Abatement Society, 1936.

———. *Ten Years' Work of Smoke Abatement in London.* [London: Coal Smoke Abatement Society, 1909].

Dickens, Charles. *Our Mutual Friend.* Edited by Michael Cotsell. Oxford: Oxford University Press, 1989.

"Discussion on 'The Beaver Report.'" *Journal of the Institute of Fuel* 28 (Apr. 1955): 175–98.

"The Domestic Smoke Nuisance." *Engineering*, 24 Nov. 1922, 653.

Doncaster, L. "Collective Inquiry as to Progressive Melanism in Lepidoptera." *Entomologist's Record and Journal of Variation* 18 (1906): 165–68, 206–54.

Douglas, C. K. M., and K. H. Stewart. "London Fog of December 5–8, 1952." *Meteorological Magazine* 82 (1953): 67–71.

Downes, Arthur, and Thomas P. Blunt. "The Influence of Light upon the Development of Bacteria." *Nature,* 12 July 1877, 218.

———. "On the Influence of Light upon Protoplasm." *Proceedings of the Royal Society of London,* 19 Dec. 1878, 199–212.

———. "Researches on the Effect of Light upon Bacteria and Other Organisms." *Proceedings of the Royal Society of London,* 6 Dec. 1877, 488–500.

"Dr. Arnott on Smokeless Fires and Pure Air in Houses." *Chambers's Journal,* 15 Sept. 1855, 174–76.

"Dust and Fog." *Engineering,* 7 Jan. 1881, 15.

"Dust and Smoke in Daily Life." *Lancet,* 14 Apr. 1906, 1054.

Dziewicki, M. H. "In Praise of London Fog." *Nineteenth Century* 26 (Dec. 1889): 1047–55.

"Editorial Notes." *Industries and Iron,* 20 Jan. 1899, 41.

Edwards, M. J. *Clean Air and Solid Fuel.* London: National Coal Board, 1970. Copy at National Society for Clean Air and Environmental Protection, Brighton.

Elderton, Ethel M. *The Relative Strength of Nurture and Nature.* London: Dulau, 1909.

"Elliott's Smoke Annihilator." *Sanitary Record,* 16 Feb. 1891, 429–30.

Ellis, Daniel. *Considerations Relative to Nuisance in Coal-Gas Works. . . .* Edinburgh: John Anderson Jr., 1828.

Engels, Friedrich. *The Condition of the Working Class in England in 1844.* 1845. Translated by W. O. Henderson and W. H. Chaloner. Stanford: Stanford University Press, 1968.

Estcourt, Charles. *Why the Air of Manchester Is So Impure.* Manchester: Manchester and Salford Noxious Vapours Abatement Association, [1888?].

Evelyn, John. *Fumifugium, or the Inconvenience of the Aer and Smoake of London Dissipated . . .* 1661. Reprinted in *The Smoake of London: Two Prophecies,* edited by James P. Lodge. Elmsford, N.Y.: Maxwell Reprint, 1969.

"Exhibition of Gas Apparatus in Leeds." *Builder,* 10 May 1879, 510.

Fayrer, Joseph. "Inaugural Address." *Journal of the Sanitary Institute* 19 (Oct. 1898): 337–59.

"Fighting the Smoke Nuisance: Factory Smoke at Sheffield—I." *Country Life,* 8 Jan. 1921, 33–34.

Fishenden, Margaret White. *The Efficiency of Low Temperature Coke in Domestic Appliances.* London: HMSO, 1921.

Fletcher, Alfred Evans. *Present Legislation on Noxious Vapours, with Suggested Improvements.* Manchester: Manchester and Salford Noxious Vapours Abatement Association, 1888.

Fletcher, B. Morley. *Report [of the] Committee for Testing Smoke-Preventing Appliances.* Manchester, 1896.

Fletcher, Herbert. *The Smoke Nuisance.* Manchester: Manchester and Salford Noxious Vapours Abatement Association, 1888.

Fletcher, Thomas. *Flame and Smoke.* Manchester: Manchester and Salford Noxious Vapours Abatement Association, 1889.

"The Fog in London." *Lancet,* 3 Jan. 1874, 27–28.

"The Fog in Parliament." *Lancet,* 5 Jan. 1889, 35.

"The Fog Question." *Chemical News,* 31 Dec. 1880, 327–28.

"Fogs." *Lancet,* 23 Oct. 1880, 665–66.

"The Fogs of London." *Nature,* 23 Dec. 1880, 165–66.

Foster, Balthazar. "Colds and Coughs." In *Birmingham Health Lectures,* 1st ser. Birmingham: Hudson, 1883.

Fothergill, John Milner. *The Town Dweller: His Needs and His Wants.* London: Lewis, 1889.

Fox, Cornelius B. *Sanitary Examinations of Water, Air, and Food.* London: J. and A. Churchill, 1878.

Fox, John Makinson. *Defective Drainage as a Cause of Disease.* Manchester: Manchester and Salford Sanitary Association, [1879].

Foxwell, G. E. "Smoke Abatement in Practice." *Journal of the Institute of Fuel* 28 (Sept. 1955): 451–57.

Frankland, Edward. "Smoke Abatement." *Nature,* 1 Mar. 1883, 407–8.

Fraser, James. "Opening Address." *Transactions of the National Association for the Promotion of Social Science,* Manchester meeting, 1879, 1–31.

Fraser-Harris, David. *National Degeneration, Being the Annual Public Lecture on the Laws of Health Delivered at the Midland Institute, 17 Sept. 1909.* Birmingham: Cornish Brothers, 1909.

"Fresh Air in the Country." *Builder,* 23 July 1859, 488.

"Frost, Fog, and Smoke." *Lancet,* 2 Jan. 1892, 40–41.

Froude, James Anthony. *Short Studies on Great Subjects.* Vol. 3. New York: Charles Scribner's Sons, 1908.

Fry, John. "Effects of a Severe Fog on a General Practice." *Lancet,* 31 Jan. 1953, 235–36.

"Further Cogitations at the Smoke Abatement Exhibition." *Builder,* 17 Dec. 1881, 745–46.

Fyfe, Peter. *Back Lands and Their Inhabitants.* Glasgow: Robert Anderson, 1901.

———. *Report of the Sanitary Department, Glasgow, for 1900.* Copy at Glasgow University Archive Services.

Galton, Douglas Strutt. *Army Sanitation: A Course of Lectures Delivered at the School of Military Engineering, Chatham.* 2nd ed. Chatham: Royal Engineers Institute, 1887.

———. *Fog and Smoke.* London, 1887.

———. *On Some Preventible Causes of Impurity in London Air.* London, 1880.

———. *On the Influences Which Principally Affect the Purity of Air in Towns.* London, 1885.

"Gas Companies and Smoke Abatement." *Builder,* 6 Oct. 1883, 470.

Glasgow Sanitary Department. *Report on the Air of Glasgow. . . ,* 1877. Copy at Royal College of Physicians and Surgeons of Glasgow.

Goodacre, William Geoffrey. *A Scheme for Drawing Off and Disposing of All Waste Gases, Impure Air, &c.* [1891]. Copy at National Society for Clean Air and Environmental Protection, Brighton.

Goodenough, Francis William. "The Fuel of the Future." In *Report of the Smoke Abatement Conference held at the Town Hall, Manchester, November 4th, 5th and 6th, 1924,* 248–54. Manchester, 1924.

Graham, John W. *The Destruction of Daylight: A Study in the Smoke Problem.* London: George Allen, 1907.

Graham, Michael. *Wage War on Smog.* London: Conservative Political Centre, 1959.

Graham, William. *Smoke Abatement in Lancashire.* Manchester: J. E. Cornish, 1896.

Grant, Madison. *The Passing of the Great Race, or The Racial Basis of European History.* New York: Charles Scribner's Sons, 1916.

Greenhow, Edward Headlam. *On the Study of Epidemic Disease, as Illustrated by the Pestilences of London.* London: T. Richards, 1858.

"The Greenwich Observatory." *Lancet,* 16 June 1906, 1704.

Harris, Frank. "The Radical Programme, III: The Housing of the Poor in Towns." *Fortnightly Review* 40 (Oct. 1883): 587–600.

Harrison, Frederic. "A Few Words about the Nineteenth Century." *Fortnightly Review* 37 (Apr. 1882): 411–26.

Hart, Ernest. "Report on the Abatement of Smoke and the Proceedings Adopted by the Smoke Abatement Committee of the National Health and Kyrle Societies." *Sanitary Record,* 15 Dec. 1881, 217–19.

Hartley, W. Noel. *Water, Air, and Disinfectants.* London: Society for Promoting Christian Knowledge, [1877].

Harvey, Alfred S. "Our Coal Supply." *Macmillan's Magazine* 26 (Sept. 1872): 375–84.

Health and Meteorology of Manchester 1 (1860). Copy in Box M3C, Manchester Medical Collection, John Rylands University Library, Manchester.

Hill, Alfred. *Report on the Health of the Borough of Birmingham*, 1876, 1882. Copies at Local Studies and History Service, Birmingham.

Hill, Octavia. "The Future of Our Commons." *Fortnightly Review* 28 (Nov. 1877): 631–41.

———. "More Air for London." *Nineteenth Century* 23 (Feb. 1888): 181–88.

Holland, Robert. *Air Pollution as Affecting Plant Life.* Manchester: Manchester and Salford Noxious Vapours Abatement Association, 1888.

"Home Fires Burning Bright." *Economist,* 27 May 1961, 901–2.

Horsfall, Thomas Coglan. *The Nuisance of Smoke from Domestic Fires, and Methods of Abating It.* Manchester: Manchester and Salford Noxious Vapours Abatement Association, [1893?].

How to Subdue Smoke: Being Popular Information on Various Practicable Means; with Comparative Results, and a Few Words about Fuel. 3rd ed. London: Effingham Wilson and Edward Stanford, 1854.

Howard, Ebenezer. *To-morrow: A Peaceful Path to Real Reform.* London: Swan Sonnenschein, 1898.

Hurst, Joseph. *English Law Relating to the Emission of Smoke from Chimneys.* London: Coal Smoke Abatement Society, [ca. 1912]. Copy at National Society for Clean Air and Environmental Protection, Brighton.

"The International Exhibition of Smoke-Preventing Appliances." *Builder,* 3 Dec. 1881, 693–707; 10 Dec. 1881, 739.

J. C. "Ozone." *Once a Week,* 14 Jan. 1865, 94–96.

Jefferies, Richard. *After London, or Wild England.* London: Cassell, 1886.

Jevons, William Stanley. *The Coal Question: An Inquiry concerning the Progress of the Nation, and the Probable Exhaustion of Our Coal-Mines.* London: Macmillan, 1865.

Johnson, James. *Change of Air, or the Pursuit of Health.* . . . London: S. Highley, 1831.

Jones, Robert. "Physical and Mental Degeneration." *Journal of the Society of Arts,* 4 Mar. 1904, 327–43.

Jones, W. Idris. "Fuel and Power Research in Britain." *Journal of the Institute of Fuel* 28 (1955): 262–71.

"Julius Berend Cohen." *Obituary Notices of Fellows of the Royal Society, 1932–35* 1 (1935): 503–13.

Kay, James Phillips. *The Moral and Physical Condition of the Working Classes Employed in the Cotton Manufacture in Manchester.* London: James Ridgway, 1832.

Kershaw, John B. C. "Notes on Recent Progress in the Campaign against Black Smoke in this Country." In Coal Smoke Abatement Society, *Papers Read at the Smoke Abatement Conferences, March 26, 27, and 28, 1912, with Discussions,* 68–75. London: Coal Smoke Abatement Society, 1912.

———. "Smoke Abatement and the New Bill." *Fortnightly Review* 120 (Dec. 1923): 963–64.

Kingzett, Charles T. *Nature's Hygiene: A Series of Essays on Popular Scientific Subjects.* . . . London: Baillière, Tindall, and Cox, 1880.

Ladies' Sanitary Association. *The Twenty-Third Annual Report.* London, 1881.

Lancet Analytical Sanitary Commission. "Report on Perfect Combustion and Smoke Prevention." *Lancet,* 21 Mar. 1891, 682–85.

———. "Second Report on Perfect Combustion and Smoke Prevention." *Lancet,* 5 Mar. 1892, 548–50.

———. "III—Gaseous Fuel, Gas-Heating, and Gas-Cooking Appliances." *Lancet,* 25 Nov. 1893, 1326–36.

Law, James. "Atmospheric Pollution," *Journal of the Royal Sanitary Institute* 5 (Nov. 1939): 202–13.

Leigh, John. *Coal-Smoke: Report to the Health and Nuisance Committees of the Corporation of Manchester.* Manchester: John Heywood, 1883.

Leslie, T. E. C. "The Known and the Unknown in the Economic World." *Fortnightly Review* 31 (June 1879): 934–49.

Littlejohn, Harvey. *Report on the Causes and Prevention of Smoke from Manufacturing Chimneys.* 1897. Copy at Sheffield Local Studies Library.

Livesey, George. "Domestic Smoke Abatement." In *Addresses, Papers, and Discussions at the Conference on Smoke Abatement, Held in London, December 12-15th, 1905,* 13-19. London: Royal Sanitary Institute, 1906.

———. *Report on the First 6 Months' Working of the Accident Fund.* South Metropolitan Gas Company, [1898]. Copy at the National Gas Archive, Warrington.

Logan, W. P. D. "Mortality in the London Fog Incident, 1952." *Lancet,* 14 Feb. 1953, 336–38.

London County Council. *Smoke Nuisance in London: Report of the Chief Officer of the Public Control Department.* 1904. Copy at London Metropolitan Archives.

"A London Fog." *Leisure Hour,* 1 Dec. 1853, 772–74.

"London Fogs." *Spectator,* 9 Jan. 1892, 45–46.

London Liberal and Reform Union. *London's Lungs.* [1896]. Copy in Hunter Papers, Box 10/1, Surrey Record Office, Woking.

Lubbock, John. "On the Preservation of Our Ancient National Monuments." *Nineteenth Century* 1 (Apr. 1877): 257–69.

Lyon, Thomas Glover. "The Air Supply of London." *Journal of Preventive Medicine* 13 (1905): 685–89.

Mackenzie, Compton. *The Vital Flame.* London: British Gas Council, 1947.

Maddock, Alfred Beaumont. *On Sydenham, Its Climate and Place, with Observations on the Efficacy of Pure Air, Especially When Combined with Intellectual and Physical Recreation in the Prevention and Treatment of Disease.* London: Simkin, Marshall, 1860.

Manchester and Salford Sanitary Association. *Annual Report, 1895, 1896.* Copies at Manchester Medical Collection, John Rylands University Library, Manchester.

"Manchester and Salford Sanitary Association: Exhibition of Sanitary Appliances." *British Architect and Northern Engineer,* 10 Aug. 1877, 67–68.

Manchester Field-Naturalists and Archaeologists' Society. *The Atmosphere of Manchester: Preliminary Report of the Air Analysis Committee.* Manchester, 1891.

"Manchester: Smoke." *Lancet,* 28 Apr. 1894, 1099–1100.

"The Manchester Smoke Abatement Exhibition." *Sanitary Record,* 15 Apr. 1882, 424–25; 15 May 1882, 475–77.

Manchester Steam Users' Association for the Prevention of Steam Boiler Explosions and for the Attainment of Economy in the Application of Steam. *A Sketch of the Foundation and of the Past Fifty Years' Activity.* Manchester: Taylor, Garnet, Evans, 1905.

Marsh, Arnold. *Smoke: The Problem of Coal and the Atmosphere.* London: Faber and Faber, [1947].

Marsh, George Perkins. *Man and Nature, or Physical Geography as Modified by Human Action.* New York: Scribner, 1864.

Maurice, C. Edmund, ed. *Life of Octavia Hill, as Told in Her Letters.* London: Macmillan, 1913.

"Meeting of the British Medical Association at Cork: Public Medicine Section." *Sanitary Record,* 15 Sept. 1879, 104–9.

"The Menace of Air Pollution." *Planning,* 16 Aug. 1954, 189–216.

Michael, W. H. "The Law in Relation to Sanitary Progress." *Sanitary Record,* 15 Feb. 1881, 281–85.

Morgan, John Edward. *The Danger of Deterioration of Race from the Too Rapid Increase of Great Cities.* London: Longmans, Green, and Co., 1866.

Morris, William. *Art and Socialism.* 2nd ed. London: Leek Bijou Reprint, [1884].

"Mr. Gladstone on Sanitary Matters." *British Architect and Northern Engineer,* 10 Aug. 1877, 70–71.

"Mr. Ruskin in the Clouds." *Builder,* 9 Feb. 1884, 190–91.

Munro, J. E. C. "Return of Rates of Wages in the Mines and Quarries in the United Kingdom with Report thereon of 1891 (C-6455)." *Economic Journal* 1, no. 4 (1891): 806–8.

Murray, Leonora. "The Miner's Home." *Nineteenth Century and After* (Jan. 1929): 9–31.

National Health Society. *Annual Report,* 1880, 1881, 1884. Copies at Radcliffe Science Library, Oxford.

"The National Health Society." *Sanitary Record,* 9 Feb. 1877, 89.

National Smoke Abatement Institution. *Smoke Abatement: Report of Council of National Smoke Abatement Institution, Submitted at a Public Meeting at the Mansion House, July 16th, 1884.* London, 1884.

"The National Smoke Abatement Institution." *Builder,* 28 July 1883, 132.

National Smoke Abatement Society. *Proceedings of the London Conference, 1943.*

———. *Smoke Abatement Exhibition Handbook and Guide.* Manchester: National Smoke Abatement Society, 1936.

———. *Smoke Abatement in Wartime.* National Smoke Abatement Society, 1940.

———. *Smoke Control Areas.* London: National Smoke Abatement Society, [ca. 1956].

———. *Smoke Prevention in Relation to Initial Post-War Reconstruction.* National Smoke Abatement Society, 1942.

National Society for Clean Air. *Notes on the Background to Clean Air. . . .* Brighton: National Society for Clean Air, [ca. 1970].

Newbigging, Thomas, and W. T. Fewtrell, eds. *King's Treatise on the Science and Practice of the Manufacture and Distribution of Coal Gas,* 3 vols. London: William B. King, 1878–82.

Nicholson, William. *Practical Smoke Prevention.* London: Sanitary Publishing, 1902.

———. *Smoke Abatement: A Manual for the Use of Manufacturers, Inspectors, Medical Officers of Health, Engineers, and Others.* London: Charles Griffin, 1905.

———. "Smoke Abatement from the Inspector's Point of View." In Coal Smoke Abatement Society, *Papers Read at the Smoke Abatement Conferences, March 26, 27, & 28, 1912, with Discussions,* 76–80. London: Coal Smoke Abatement Society, 1912.

"No Headaches in Gas Works." *Gas Bulletin* 22 (1933): 57.

"Notes." *Nature,* 2 Apr. 1885, 513.

"Notes of the Sanitary Congress at York." *Builder,* 2 Oct. 1886, 499–502.

Notter, J. Lane. "Sanitary Notes." *Sanitary Record,* 15 June 1880, 445–49.

"The Obelisk." *Builder,* 12 Oct. 1878, 1074.

"Observations in a London Fog." *Hogg's Instructor* 5 (1855): 53–55.

Oliver, F. W. "On the Effects of Urban Fog upon Cultivated Plants." *Journal of the Royal Horticultural Society* 13 (1891): 139–51.

Oliver, Thomas. "Coke-Men and By-Products Workers: Their Complaints and Maladies." *British Medical Journal,* 31 May 1930, 992–94.

"The Opening of the Smoke Abatement Exhibition at South Kensington." *Sanitary Record,* 15 Dec. 1881, 227–60.

Oswald, Neville C., James T. Harold, and W. J. Martin. "Clinical Pattern of Chronic Bronchitis." *Lancet,* 26 Sept. 1953, 639–43.

Page, William, ed. *Commerce and Industry: Tables of Statistics for the British Empire from 1815.* London: Constable, 1919.

Palm, Theobald A. "The Geographical Distribution and Ætiology of Rickets." *Practititioner* 45 (July–Dec. 1890): 270–79, 321–42.

Parkes, Louis C. "The Air and Water of London: Are They Deteriorating?" *Transactions of the Sanitary Institute of Great Britain* 13 (1892): 59–69.

Percy, George. "Coal and Smoke." *Quarterly Review* 119 (Apr. 1866): 435–72.

Phelps, Earle B. "Stream Pollution by Industrial Wastes and Its Control." In *A Half Century of Public Health,* edited by Mazÿck P. Ravenel, 197–208. New York: American Public Health Association, 1921.

Playfair, Lyon. "Pure Air, Pure Water, and Pure Soil." *Builder,* 10 Oct. 1874, 846.

Pollock, William Frederick. "Smoke Prevention." *Nineteenth Century* 9 (Mar. 1881): 478–90.

Poore, George Vivian. "Light, Air, and Fog." *Transactions of the Sanitary Institute of Great Britain* 14 (1893): 32–36.

Popplewell, William Charles. *The Prevention of Smoke, Combined with the Economical Combustion of Fuel.* London: Scott, Greenwood, 1901.

Pownall, A. E. *Some Considerations of the Effects of Air Pollution on Health.* Air Pollution Lectures. Manchester: Manchester and Salford Noxious Vapours Abatement Association, [1892].

Preece, William Henry. "Gas versus Electricity." *Nature,* 23 Jan. 1879, 261–62.

Price, Richard. *A Supplement to the Second Edition of the Treatise.* London: T. Cadell, 1772.

Pritchard, Richard. "The Influence of Ventilation on the Type of the Disease." *Public Health* 15 (1902–3): 385–93.

"Proceedings of the Health Section." *Sanitary Record,* 20 Oct. 1879, 138–41.

"Public Monuments." *Edinburgh Review* 115 (Apr. 1862): 541–65.

"Pure Air and Noxious Vapours." *Builder,* 26 June 1880, 785.

Ransome, Arthur. *Foul Air and Lung Disease.* Manchester: Manchester and Salford Sanitary Association, [ca. 1877].

"The Reign of Darkness." *Spectator,* 19 Jan. 1889, 85–86.

"Report on the Decay of Stone at Westminster." *Builder,* 5 Oct. 1861, 677–80.

Review of *Degeneration: A Chapter in Darwinism,* by E. Ray Lankester. *London Quarterly Review* 56, no 112. (1881): 353–66.

Review of *London Smoke and Fog,* by Frederick Edwards. *Chemical News,* 7 Jan. 1881, 10–11.

Review of *The Sanitary Inspector's Hand-Book,* by Albert Taylor. *Builder,* 7 July 1900, 10.

Reynolds, F. *The Law Relating to Air Pollution.* Brighton: National Society for Clean Air, n.d.

Richardson, Benjamin Ward. *A Curse of Civilisation.* London, 1913.

———."Health and Civilisation." *Journal of the Society of Arts,* 15 Oct. 1875, 948–54.

———. "On Ozone in Relation to Health and Disease." *Popular Science Review* 5, no. 18 (1866): 29–40.

———. *On the Poisons of the Spreading Diseases: Their Nature and Mode of Distribution.* London: John Churchill, 1867.

Richmond, William Blake. "The Black City." *Pall Mall Magazine* 23 (Apr. 1901): 462–73.

———. "The Smoke Plague of London." In *London of the Future,* edited by Aston Webb, 261–70. 1921.

Roberts, Robert. *The Classic Slum: Salford Life in the First Quarter of the Century.* Manchester: Manchester University Press, 1971.

———. *A Ragged Schooling: Growing Up in the Classic Slum.* Manchester: Mandolin, 1984.

Ross, Owen C. D. "A Cure for London Fogs." *Gentleman's Magazine* 274 (Mar. 1893): 228–36.

Ruskin, John. "The Lamp of Memory." 1849. In *The Seven Lamps of Architecture.* 2nd ed. New York: John Wiley, 1890.

Russell, Francis Albert Rollo. *The Atmosphere in Relation to Human Life and Health.* Washington, D.C.: Smithsonian Institution, 1896.

———. "Haze, Fog, and Visibility." *Quarterly Journal of the Royal Meteorological Society* 23 (1897): 10–24.

———. *London Fogs.* London: Edward Stanford, 1880.

———. *Smoke in Relation to Fogs in London.* London, 1888.

Russell, James Burn. *An Address Delivered at the Opening of the Section of Public Medicine, at the Annual Meeting of the British Medical Association in Sheffield, August 1876.* Glasgow: Robert Anderson, 1876.

———. *Public Health Administration in Glasgow: A Memorial Volume of the Writings of James Burn Russell.* Edited by Archibald Kerr Chalmers. Glasgow: James Maclehose, 1905.

Salford Health Department. *Smoke Control Means Clean Air for You and Your Family.* n.d. Copy at National Society for Clean Air and Environmental Protection, Brighton.

"The Sanitary Inspectors' Association: Conference at Lincoln." *Builder,* 12 Aug. 1899, 155–57.

"Sanitary Inspectors' Association: Sixteenth Annual Dinner." *Builder,* 11 Feb. 1899, 147–48.

"The Sanitary Institute Congress at Leeds." *Builder,* 18 Sept. 1897, 222–23.

Scott, Fred. The Administration of the Law Relating to Smoke. [1895?]. Copy at National Society for Clean Air and Environmental Protection, Brighton.

Scott, Robert H. "Fifteen Years' Fogs in the British Islands, 1876–1890," *Quarterly Journal of the Royal Meteorological Society* 19 (1893): 229–38.

———. "London Fogs." *Longman's Magazine* 9 (Apr. 1887): 607–14.

Seddon, John P. "The Warming of Houses." *Architect,* 14 Dec. 1872, 327–29.

Shaw, William Napier. "The Treatment of Smoke: A Sanitary Parallel." *Journal of the Sanitary Institute* 23, no. 3 (1902): 318–34.

Shaw, William Napier, and John Switzer Owens. *The Smoke Problem of Great Cities.* London: Constable, 1925.

Siemens, Charles William. *Science and Industry.* Birmingham, [1881].

Silverthorne, Arthur. *The Purchase of Gas and Water Works, with the Latest Statistics of Municipal Gas and Water Supply.* London: Crosby Lockwood, 1881.

Simon, John. *Public Health Reports.* 2 vols. Edited by Edward Seaton. London: Sanitary Institute of Great Britain, 1887.

Simpson, Thomas Bartlett. *Gas-Works: The Evils Inseparable from Their Existence in Populous Places. . . .* London: William Freeman, 1866.

Sims, George R. *How the Poor Live.* 1883. Reprinted in *How the Poor Live, and Horrible London.* London: Chatto and Windus, 1889.

"Sir William B. Richmond, R.A., K.C.B., on the Smoke Nuisance." *London Argus,* 14 Jan. 1899, 217–18.

Smith, Robert Angus. *A Centenary of Science in Manchester: For the Hundredth Year of the Literary and Philosophical Society of Manchester (1881).* London: Taylor and Francis, 1883.

———. "On the Air of Towns." *Quarterly Journal of the Chemical Society* 11 (1859): 196–235.

———. *On Some Invisible Agents of Health and Disease.* Manchester: Manchester and Salford Sanitary Association, 1878.

———. "On the Estimation of the Organic Matter of the Air." *Proceedings of the Royal Institution of Great Britain* 3 (1859): 89–94.

Smith, Thomas Southwood. *The Common Nature of Epidemics and Their Relation to Climate and Civilization; also Remarks on Contagion and Quarantine.* Edited by T. Baker. London: N. Trübner, 1866.

Smith, William B. "Should the Domestic Smoke Nuisance Be Any Longer Tolerated?" In Coal Smoke Abatement Society, *Papers Read at the Smoke Abatement Conferences, March 26, 27, and 28, 1912, with Discussions,* 62–67. London: Coal Smoke Abatement Society, 1912.

Smith, William R. *The Laboratory Text-Book of Public Health.* London: Henry Renshaw, 1896.

"Smoke." *Engineer,* 9 Dec. 1898, 569–70.

"Smoke Abatement." *Builder,* 10 Nov. 1883, 635.

"Smoke Abatement." *Engineering,* 7 Aug. 1896, 178–79.

"Smoke Abatement." *Gas Bulletin* 11 (July 1922): 141.

"Smoke Abatement." *Industries,* 7 Nov. 1890, 454.

"Smoke Abatement." *Nature,* 19 July 1883, 278–79.

Smoke Abatement Committee. *Report of the Smoke Abatement Committee, 1882.* London: Smith, Elder, 1883.

"The Smoke-Abatement Exhibition." *Lancet,* 17 Dec. 1881, 1068–69; 24 Dec. 1881, 1090.

"Smoke-Abatement Exhibition." *Nature,* 8 Dec. 1881, 121–22.

"The Smoke Abatement Exhibition." *Nature,* 5 Jan. 1882, 219–21.

"Smoke Abatement Exhibition at South Kensington." *Sanitary Record,* 15 Feb. 1882, 339–40.

"The Smoke Abatement Institute." *Nature,* 17 Aug. 1882, 371.

Smoke Abatement League of Great Britain. *First Annual Report,* 1909–10.

———. *Report of the Smoke Abatement Conference held at the Town Hall, Manchester, November 4th, 5th and 6th, 1924.* Manchester, 1924.

———. *Report of the Smoke Abatement Exhibition Held at Bingley Hall, Birmingham, September 7th–10th, 1926.* [1926]. Copy at Local Studies and History Service, Birmingham.

———. *The Universal Smoke Abatement Exhibition, Bingley Hall, Birmingham, September 6th–18th, 1926: Official Guide and Catalogue.* [1926]. Copy at Local Studies and History Service, Birmingham.

"Smoke-Abating Appliances." *Lancet,* 3 Dec. 1881, 970.

"Smoke into Sales." *Gas Bulletin* 28 (Feb. 1939): 32.

"The Smoke Nuisance." *Builder,* 31 Dec. 1881, 835.

"The Smoke Nuisance." *Builder,* 12 Aug. 1899, 143–45.

"Smoke Nuisance Abatement." *Lancet,* 8 Oct. 1887, 722–23.

"Smoke Prevention at Gas Works (A New Process of Gas Making)." *Lancet,* 12 Mar. 1904, 746.

"Smokeless Heat." *Builder,* 10 Dec. 1881, 715.

Society for the Protection of Ancient Buildings. *Annual Report.* 1882.

Sohncke, Leonhard. "The Problem of the Exhaustion of Coal." *Open Court* 4 (1890): 2375–76, 2389–92.

Spence, Peter. *Coal, Smoke, and Sewage, Scientifically and Practically Considered.* Manchester, 1857.

———. "Smoke and Fog." *Chemical News,* 25 Mar. 1881, 142–43.

Steer, W. R. Hornby. *The Law of Smoke Nuisances.* London: National Smoke Abatement Society, 1938.

Stone, W. H. "On Fog." *Popular Science Review,* n.s., 5, no. 17 (1881): 27–39.

Stott, Benjamin. *Songs for the Millions and Other Poems.* Middleton: W. Horsman, 1843.

T., H. "Dr. Mohr on 'The Atmosphere and Our Obelisk.'" *Builder,* 9 Mar. 1878, 251.

Tait, Lawson. *Gas as Fuel.* Birmingham: Hudson, n.d.

Taylor, J. S. *Smoke and Health.* Manchester: National Smoke Abatement Society, 1929.

Thomson, George Carruthers. *On Smoke Abatement with Reference to Steam Boiler Furnaces, Read before the Philosophical Society of Glasgow, 6th February 1895.* Reprinted, Calcutta Boiler Commission, n.d.

Thorne, Will. *My Life's Battles.* London: George Newnes, [1925].

Thwaite, B. H. "London Fog: A Scheme to Abolish It." *National Review* 20 (Nov. 1892): 360–67.

Tocqueville, Alexis de. *Journeys to England and Ireland.* 1835. Translated by George Lawrence and K. P. Mayer, and edited by J. P. Mayer. New Brunswick, N.J.: Transaction Press, 1988.

"The Toll of Fog." *British Medical Journal,* 7 Feb. 1953, 321.

Underhill, C. B. "Recollections of Life at an Open-Air Sanatorium." *Good Words and Sunday Magazine* 44 (1903): 183–86.

United Nations. *Demographic Yearbook, 1954.* New York: United Nations, 1954.

"The Ventilation of the House of Commons." *Sanitary Record,* 16 Feb. 1877, 97–99.

Voelcker, Augustus. "On the Injurious Effects of Smoke on Certain Building Stones and on Vegetation." *Journal of the Society of Arts,* 22 Jan. 1864, 146–53.

Wallace, William. "The Air of Towns." In *Lectures on the Theory and General Prevention and Control of Infectious Diseases, and on Water Supply, Sewage Disposal, and Food,* by J. B. Russell and William Wallace. Glasgow: Robert Anderson, 1879.

Walters, F. Rufenacht. *Sanatoria for Consumptives: A Critical and Detailed Description.* 3rd ed. London: Swan Sonnenschein, 1905.

Webber, William Hosgood Young. *Town Gas and Its Uses for the Production of Light, Heat, and Motive Power.* London: Constable, 1907.

Webster, A. D. *Town Planting and the Trees, Shrubs, Herbaceous and Other Plants That Are Best Adapted for Resisting Smoke.* London: George Routledge, [1910].

Wells, H. G. *The Time Machine.* 1895. Reprinted in H. G. Wells, *The Time Machine, The Wonderful Visit, and Other Stories.* New York: Charles Scribner's Sons, 1924. Vol. 1 of the Atlantic Edition of H. G. Wells.

White, F. Buchanan. "Melanism, &c., in Lepidoptera." *Entomologist* 10 (May 1877): 126–29.

White, J. *On Health, as Depending on the Condition of Air. . . .* London: Hamilton, Adams, 1859.

Wilkins, E. T. "Air Pollution and the London Fog of December, 1952." *Journal of the Royal Sanitary Institute* 74 (Jan. 1954): 1–21.

Wilson, Albert. "The Great Smoke-Cloud of the North of England and Its Influence on Plants." Abstract in the *Report of the Seventieth Meeting of the British Association for the Advancement of Science Held in Bradford September 1900,* 930–31 London: John Murray, 1900.

Winkler, Clemens. "The Influence of the Combustion of Coal upon our Atmosphere." *Open Court* 1 (May 1887): 197–99.

Wyld, George. *Notes of My Life.* London: Kegan, 1903.

Secondary Sources

Theses and Dissertations

Blackden, Stephanie M. "The Development of Public Health Administration in Glasgow, 1842–1872." Ph.D. thesis, Edinburgh University, 1976.

Harrison, Michael. "Social Reform in Late Victorian and Edwardian Manchester, with Special Reference to T. C. Horsfall." Ph.D. thesis, University of Manchester, 1988.

Mills, Mary. "The Early Gas Industry and Its Residual Products in East London." Ph.D. thesis, Open University, 1995.

Ryan, P. A. "Public Health and Voluntary Effort in Nineteenth-Century Manchester, with Particular Reference to the Manchester and Salford Sanitary Association." Master's thesis, University of Manchester, [1974].

Books and Articles

Aalen, F. H. A. "Lord Meath, City Improvement, and Social Imperialism." *Planning Perspectives* 4, no. 2 (1989): 127–52.

Adams, Annmarie. *Architecture in the Family Way: Doctors, Houses, and Women, 1870–1900.* Montreal and Kingston: McGill–Queen's University Press, 1996.

Ashby, Eric, and Mary Anderson. *The Politics of Clean Air.* Monographs on Science, Technology, and Society. Oxford: Clarendon Press, 1981.

Ashton, John, and Howard Seymour. *The New Public Health: The Liverpool Experience.* Milton Keynes: Open University Press, 1988; reprinted 1995.

Ashworth, William, and Mark Pegg. *The History of the British Coal Industry.* Vol. 5, *1946–1982: The Nationalized Industry.* Oxford: Clarendon Press, 1986.

Bailey, Maurice H. "A Birmingham Clean Air Pioneer: The Influence of John Cadbury." *Smokeless Air* 102 (Summer 1957): 260–62.

Baldwin, Peter C. "How Night Air Became Good Air, 1776–1930." *Environmental History* 8 (July 2003): 412–29.

Barnes, Barry, and Steven Shapin, eds. *Natural Order: Historical Studies of Scientific Culture.* Beverly Hills: Sage, 1979.

Bartrip, P. W. J. *Mirror of Medicine: A History of the British Medical Journal.* Oxford: Clarendon Press, 1990.

Barty-King, Hugh. *New Flame: How Gas Changed the Commercial, Domestic and Industrial Life of Britain between 1813 and 1984.* Tavistock, England: Graphmitre, 1984.

Behlmer, George K. "Ernest Hart and the Social Thrust of Victorian Medicine." *British Medical Journal,* 3 Oct. 1990, 711–13.

Bell, Enid Moberly. *Octavia Hill: A Biography.* 2nd ed. London: Constable, 1943.

Bell, Michelle L., and Devra Lee Davis. "Reassessment of the Lethal London Fog of 1952: Novel Indicators of Acute and Chronic Consequences of Acute Exposure to Air Pollution." *Environmental Health Perspectives* 109, supp. 3 (June 2001): 389–94.

Bernstein, Henry T. "The Mysterious Disappearance of Edwardian London Fog." *London Journal* 1, no. 2 (1975): 189–206.

Boon, Timothy. "'The Smoke Menace': Cinema, Sponsorship and the Social Relations of Science in 1937." In *Science and Nature: Essays in the History of the Environmental Sciences,* edited by Michael Shortland, 57–88. [Stanford in the Vale]: British Society for the History of Science, 1993.

Bouman, Mark J. "The 'Good Lamp is the Best Police' Metaphor and Ideologies of the Nineteenth-Century Urban Landscape." *American Studies* 32 (Spring 1991): 63–78.

Boyd, Nancy. *Three Victorian Women Who Changed Their World: Josephine Butler, Octavia Hill, Florence Nightingale.* New York: Oxford University Press, 1982.

Brears, Peter. *Images of Leeds, 1850–1960.* Derby: Breedon Books, 1992.

Brenner, Joel Franklin. "Nuisance Law and the Industrial Revolution." *Journal of Legal Studies* 3 (June 1974): 403–33.

Briggs, Asa. *Victorian Cities.* 1963. Reprint, Berkeley: University of California Press, 1993.

Brimblecombe, Peter. *The Big Smoke: A History of Air Pollution in London since Medieval Times.* London: Methuen, 1987.

———. "London Air Pollution, 1500–1900." *Atmospheric Environment* 11, no. 12 (1977): 1157–62.

———. "Long Term Trends in London Fog." *Science of the Total Environment* 22, no. 1 (1981): 19–29.

Brock, William H. *The Chemical Tree: A History of Chemistry.* New York: W. W. Norton, 2000.

———. *Justus von Liebig: The Chemical Gatekeeper.* Cambridge: Cambridge University Press, 1997.

Burgess, William A. *Recognition of Health Hazards in Industry: A Review of Materials and Processes.* New York: John Wiley and Sons, 1981.

Byers, William, Martin B. Meyers, and Donna E. Mooney. "Analysis of Soil from a Disused Gasworks." *Water, Air, and Soil Pollution* 73, nos. 1–4 (1994): 1–9.

Bynum, W. F. "Policing Hearts of Darkness: Aspects of the International Sanitary Conferences." *History and Philosophy of the Life Sciences* 15, no. 3 (1993): 421–34.

———. *Science and the Practice of Medicine in the Nineteenth Century.* Cambridge: Cambridge University Press, 1994.

Chamberlin, J. Edward. "An Anatomy of Cultural Melancholy." *Journal of the History of Ideas* 42, no. 4 (1981): 691–705.

Chamberlin, J. Edward, and Sander L. Gilman, eds. *Degeneration: The Dark Side of Progress.* New York: Columbia University Press, 1985.

Chandler, Tertius, and Gerald Fox. *3000 Years of Urban Growth.* New York: Academic Press, 1974.

Chantler, Philip. *The British Gas Industry: An Economic Study.* Manchester: Manchester University Press, 1938.

Charlton, Christopher. "The National Health Society Almanack, 1883." *Local Population Studies* 32 (Spring 1984): 54–57.

Cherfas, Jeremy. "Clean Air Revives the Peppered Moth." *New Scientist,* 2 Jan. 1986, 17.

Church, Roy, with the assistance of Alan Hall and John Kanefsky. *The History of the British Coal Industry.* Vol. 3, *1830–1913: Victorian Pre-Eminence.* Oxford: Clarendon Press, 1986.

Clapp, B. W. *An Environmental History of Britain since the Industrial Revolution.* New York: Longman, 1994.

Coles, Nicholas. "Sinners in the Hands of an Angry Utilitarian: J. P. Kay (-Shuttleworth), *The Moral and Physical Condition of the Working Classes in Manchester* (1832)." *Bulletin of Research in the Humanities* 86, no. 4 (1985): 453–88.

Cook, Chris, and John Stevenson. *The Longman Handbook of Modern British History, 1714–2001.* 4th ed. London: Longman, 2001.

Cook, E. T. *The Life of John Ruskin.* 2nd ed. 2 vols. London: George Allen, 1912.

Cooter, Roger. "Anticontagionism and History's Medical Record." In *The Problem of Medical Knowledge: Examining the Social Construction of Medicine,* edited by Peter Wright and Andrew Treacher, 87–108. Edinburgh: Edinburgh University Press, 1982.

———. "The Power of the Body: The Early Nineteenth Century." In *Natural Order: Historical Studies of Scientific Culture,* edited by Barry Barnes and Steven Shapin, 73–92. Beverly Hills: Sage, 1979.

Coppock, J. T., and Hugh C. Prince, eds. *Greater London.* London: Faber and Faber, 1964.

Corbin, Alain. *The Foul and the Fragrant: Odor and the French Social Imagination.* Translated by Miriam L. Kochan. Cambridge: Harvard University Press, 1986.

Cosgrove, Denis, and John E. Thornes. "Of Truth of Clouds: John Ruskin and the Moral Order in Landscape." In *Humanistic Geography and Literature,* edited by Douglas C. D. Pocock, 20–46. London: Croom Helm, 1981.

Cotterill, M. S. "The Development of Scottish Gas Technology, 1817–1914: Inspiration and Motivation." *Industrial Archaeology Review* 5, no. 1 (1980–81): 19–40.

Craxford, S. R., and Marie-Louise P. M. Weatherley. "Air Pollution in Towns in the United Kingdom." *Philosophical Transactions of the Royal Society of*

London. Series A, Mathematical and Physical Sciences 269, no. 1199 (May 1971): 503–13.

Cronon, William. *Nature's Metropolis: Chicago and the Great West.* New York: W. W. Norton, 1991.

Cumbler, John T. "Whatever Happened to Industrial Waste? Reform, Compromise, and Science in Nineteenth Century Southern New England." *Journal of Social History* 29 (Fall 1995): 149–71.

Danahay, Martin A. "Matter Out of Place: The Politics of Pollution in Ruskin and Turner." *Clio* 21:1 (1991): 61–77.

Darley, Gillian. *Octavia Hill.* London: Constable, 1990.

Davis, Devra Lee. *When Smoke Ran Like Water: Tales of Environmental Deception and the Battle against Pollution.* New York: Basic Books, 2002.

Davis, John. "Modern London, 1850–1939." *London Journal* 20, no. 2 (1995): 56–90.

Davison, Graeme. "The City as a Natural System: Theories of Urban Society in Early Nineteenth-Century Britain." In *The Pursuit of Urban History,* edited by Derek Fraser and Anthony Sutcliffe, 349–70. London: Edward Arnold, 1983.

Dellheim, Charles. *The Face of the Past: The Preservation of the Medieval Inheritance in Victorian England.* Cambridge: Cambridge University Press, 1982.

Dennis, Norman, Fernando Henriques, and Clifford Slaughter. *Coal Is Our Life: An Analysis of a Yorkshire Mining Community.* London: Eyre and Spottiswoode, 1956.

Dingle, A. E. "'The Monster Nuisance of All': Landowners, Alkali Manufacturers, and Air Pollution, 1828–64." *Economic History Review,* 2nd ser., 35, no. 4 (1982): 529–48.

Dobson, Mary J. *Contours of Death and Disease in Early Modern England.* Cambridge: Cambridge University Press, 1997.

Douglas, Mary. *Purity and Danger: An Analysis of Concepts of Pollution and Taboo.* London: Routledge and Kegan Paul, 1966.

Dow, J. C. R. *The Management of the British Economy, 1945–60.* Cambridge: Cambridge University Press, 1964.

Dyos, H. J., and Michael Wolff, eds. *The Victorian City: Images and Realities,* 2 vols. London: Routledge and Kegan Paul, 1973.

Evans, David. *A History of Nature Conservation in Britain.* London: Routledge, 1992.

Eyler, John M. "The Conversion of Angus Smith: The Changing Role of Chemistry and Biology in Sanitary Science, 1850–1880." *Bulletin of the History of Medicine* 54 (1980): 216–34.

———. *Victorian Social Medicine: The Ideas and Methods of William Farr.* Baltimore: Johns Hopkins University Press, 1979.

Falkus, M. E. "The British Gas Industry before 1850." *Economic History Review*, 2d ser., 20, no. 3 (1967): 494–508.

Fawcett, Jane. "A Restoration Tragedy: Cathedrals in the Eighteenth and Nineteenth Centuries." In *The Future of the Past: Attitudes to Conservation, 1174–1974*, edited by Jane Fawcett. London: Thames and Hudson, 1976.

Fitch, Raymond E. *The Poison Sky: Myth and Apocalypse in Ruskin.* Athens: Ohio University Press, 1982.

Fitter, R. S. R. *London's Natural History.* London: Collins, 1945.

Flick, Carlos. "The Movement for Smoke Abatement in 19th-Century Britain." *Technology and Culture* 21 (Jan. 1980): 29–50.

Flinn, Michael W., with the assistance of David Stoker. *The History of the British Coal Industry.* Vol. 2, *1700–1830: The Industrial Revolution.* Oxford: Clarendon Press, 1984.

Francis, Huw. "Understanding Medical Officers of Health." *Public Health* 102 (1988): 545–53.

Fraser, Derek, and Anthony Sutcliffe, eds. *The Pursuit of Urban History.* London: Edward Arnold, 1983.

Gaskell, S. Martin. "Gardens for the Working Class: Victorian Practical Pleasure." *Victorian Studies* 23 (Summer 1980): 479–501.

Gates, Barbara T. *Kindred Nature: Victorian and Edwardian Women Embrace the Living World.* Chicago: University of Chicago Press, 1998.

Gaze, John. *Figures in a Landscape: A History of the National Trust.* N.p.: Barrie and Jenkins, in association with the National Trust, 1988.

Gibbon, Gwilym, and Reginald W. Bell. *History of the London County Council, 1889–1939.* London: Macmillan, 1939.

Gilbert, Bentley B. *The Evolution of National Insurance in Great Britain: The Origins of the Welfare State.* London: Michael Joseph, 1966.

———. "Health and Politics: The British Physical Deterioration Report of 1904." *Bulletin of the History of Medicine* 39, no. 2 (Mar.–Apr. 1965): 143–53.

Goldberger, Paul. "The Big Top." *New Yorker,* 27 Apr./4 May 1998, 152–59.

Goldman, Lawrence. "Debate: Reply." *Past and Present* 121 (Nov. 1988): 214–19.

———. "The Social Science Association, 1857–1886: A Context for Mid-Victorian Liberalism." *English Historical Review* 101 (1986): 95–134.

Greenslade, William. Review of *The Degeneracy Crisis and Victorian Youth,* by Thomas E. Jordan. *Victorian Studies* 38, no. 2 (1995): 273–74.

Gugliotta, Angela. "'Dr. Sharp with His Little Knife': Therapeutic and Punitive Origins of Eugenic Vasectomy: Indiana, 1892–1921." *Journal of the History of Medicine and Allied Sciences* 53, no. 4 (1998): 371–406.

Guha, Ramachandra. *Environmentalism: A Global History.* New York: Longman, 2000.

Hamlin, Christopher. "Environmental Sensibility in Edinburgh, 1839–1840: The 'Fetid Irrigation' Controversy." *Journal of Urban History* 20 (May 1994): 311–39.

———. "Predisposing Causes and Public Health in Early Nineteenth-Century Medical Thought." *Social History of Medicine* 5 (Apr. 1992): 43–70.

———. "Providence and Putrefaction: Victorian Sanitarians and the Natural Theology of Health and Disease." *Victorian Studies* 28 (Spring 1985): 381–411.

———. *A Science of Impurity: Water Analysis in Nineteenth-Century Britain.* Berkeley and Los Angeles: University of California Press, 1990.

Hannah, Leslie. *Electricity before Nationalisation: A Study of the Development of the Electricity Supply Industry in Britain to 1948.* Baltimore: Johns Hopkins University Press, 1979.

Hannaway, Caroline. "Environment and Miasmata." In *Companion Encyclopedia of the History of Medicine,* edited by W. F. Bynum and Roy Porter, 1:292–308. 2 vols. London: Routledge, 1993.

Harrison, Brian. *Peaceable Kingdom: Stability and Change in Modern Britain.* Oxford: Clarendon Press, 1982.

Harrison, Mark. "Tropical Medicine in Nineteenth-Century India." *British Journal for the History of Science* 25, no. 3 (1992): 299–318.

Harrison, Michael. "Thomas Coglan Horsfall and 'the Example of Germany.'" *Planning Perspectives* 6, no. 3 (1991): 297–314.

Harrison, Peter. "Subduing the Earth: Genesis 1, Early Modern Science, and the Exploitation of Nature." *Journal of Religion* 79, no. 1 (1999): 86–109.

Harrison, Roy M., ed. *Pollution: Causes, Effects, and Control.* 2nd ed. Cambridge: Royal Society of Chemistry, 1990.

Hasian, Marouf Arif. *The Rhetoric of Eugenics in Anglo-American Thought.* Athens: University of Georgia Press, 1996.

Hawes, Richard. "The Control of Alkali Pollution in St. Helens, 1862–1890." *Environment and History* 1, no. 2 (1995): 159–71.

Hawkins, Mike. *Social Darwinism in European and American Thought, 1860–1945: Nature as a Model and Nature as a Threat.* Cambridge: Cambridge University Press, 1997.

Headrick, Daniel. *The Tools of Empire: Technology and European Imperialism in the Nineteenth Century.* New York: Oxford University Press, 1981.

Hicks, Joe, and Grahame Allen. *A Century of Change: Trends in UK Statistics since 1900.* London: House of Commons Library, 1999.

Hill, William Thomson. *Octavia Hill: Pioneer of the National Trust and Housing Reformer.* London: Hutchinson, 1956.

Hills, Richard L. *Power from Steam: A History of the Stationary Steam Engine.* New York: Cambridge University Press, 1989.

Holmes, Frederic L. "Elementary Analysis and the Origins of Physiological Chemistry." *Isis* 54, no. 1 (1963): 50–81.

Hoolihan, Christopher. "Health and Travel in Nineteenth-Century Rome." *Journal of the History of Medicine and the Allied Sciences* 44, no. 4 (1989): 462–85.

House, John. *Monet: Nature into Art.* New Haven: Yale University Press, 1986.

Howe, G. Melvyn. *People, Environment, Disease and Death: A Medical Geography of Britain throughout the Ages.* Cardiff: University of Wales Press, 1997.

Huch, Ronald K. "The National Association for the Promotion of Social Science: Its Contribution to Victorian Health Reform, 1857–1886." *Albion* 17 (Fall 1985): 279–300.

Humber, Sean. *Gas Works Sites in London: An Investigation into Contaminated Land.* London: Friends of the Earth, 1991.

Hurley, Andrew. *Environmental Inequalities: Class, Race, and Industrial Pollution in Gary, Indiana, 1945–1980.* Chapel Hill: University of North Carolina Press, 1995.

Huxley, Gervas. *Victorian Duke: The Life of Hugh Lupus Grosvenor, First Duke of Westminster.* London: Oxford University Press, 1967.

Inglis, K. S. *Churches and the Working Classes in Victorian England.* London: Routledge and Kegan Paul, 1963.

Intergovernmental Panel on Climate Change. *Climate Change: The IPCC Scientific Assessment,* edited by J. T. Houghton, G. J. Jenkins, and J. J. Ephraums. New York: Cambridge University Press, 1990.

Jacobsen, Mark Z. *Atmospheric Pollution: History, Science, and Regulation.* New York: Cambridge University Press, 2002.

Jarcho, Saul. "A Cartographic and Literary Study of the Word *Malaria.*" *Journal of the History of Medicine and the Allied Sciences* 25, no. 1 (1970): 31–39.

Jenkins, Jennifer, and Patrick James. *From Acorn to Oak Tree: The Growth of the National Trust, 1895–1994.* London: Macmillan, 1994.

Jenner, Mark. "The Politics of London Air: John Evelyn's *Fumifugium* and the Restoration." *Historical Journal* 38, no. 3 (1995): 535–51.

Jolly, W. P. *Sir Oliver Lodge.* London: Constable, 1974.

Jones, Greta. *Social Darwinism and English Thought: The Interaction between Biological and Social Theory.* Brighton: Harvester, 1980.

Jordan, Thomas E. *The Degeneracy Crisis and Victorian Youth.* Albany: State University of New York Press, 1993.

Jordanova, Ludmilla. "The Social Construction of Medical Knowledge." *Social History of Medicine* 8 (Dec. 1995): 361–81.

Kargon, Robert H. *Science in Victorian Manchester: Enterprise and Expertise.* Baltimore: Johns Hopkins University Press, 1977.

Kearns, Gerry. "Biology, Class and the Urban Penalty." In *Urbanising Britain: Essays on Class and Community in the Nineteenth Century*, edited by Gerry Kearns and W. J. Withers, 12–30. Cambridge: Cambridge University Press, 1991.

Klinenberg, Eric. *Heat Wave: A Social Autopsy of Disaster in Chicago*. Chicago: University of Chicago Press, 2002.

Lambert, Royston. *Sir John Simon, 1816–1904, and English Social Administration*. London: MacGibbon and Kee, 1963.

LeCain, Timothy. "The Limits of 'Eco-Efficiency': Arsenic Pollution and the Cottrell Electrical Precipitator in the U.S. Copper Smelting Industry." *Environmental History* 5, no. 3 (2000): 336–51.

Lees, D. R., E. R. Creed, and J. G. Duckett. "Atmospheric Pollution and Industrial Melanism." *Heredity* 30 (1973): 227–32.

Levine, Steven Z. *Monet and His Critics*. New York: Garland, 1976.

Lewis, Jane. *Women and Social Action in Victorian and Edwardian England*. Stanford: Stanford University Press, 1991.

Lowenthal, David. *The Past Is a Foreign Country*. Cambridge: Cambridge University Press, 1985, reprinted 1990.

Luckin, Bill. "'The Heart and Home of Horror': The Great London Fogs of the Late Nineteenth Century." *Social History* 28 (Jan. 2003): 31–48.

———. *Pollution and Control: A Social History of the Thames in the Nineteenth Century*. Bristol: Adam Hilger, 1986.

———. "Pollution in the City." In *The Cambridge Urban History of Britain*. Vol. 3, *1840–1950*, edited by Martin Daunton, 207–28. Cambridge: Cambridge University Press, 2000.

McCormick, Brian Joseph. *Industrial Relations in the Coal Industry*. London: Macmillan, 1979.

McCrone, Kathleen E. "The National Association for the Promotion of Social Science and the Advancement of Victorian Women." *Atlantis* 8 (Fall 1982): 44–66.

McEvoy, Arthur F. "Working Environments: An Ecological Approach to Industrial Health and Safety." *Technology and Culture* 36, no. 2, supp. (1995): S145–72.

MacGregor, Alexander. *Public Health in Glasgow, 1905–1946*. Edinburgh: E. and S. Livingstone, 1967.

MacLeod, Roy M. "The Alkali Acts Administration, 1863–84: The Emergence of the Civil Scientist." *Victorian Studies* 9 (Dec. 1965): 83–112.

MacLeod, Roy M., and E. Kay Andrews. "The Origins of the D.S.I.R.: Reflections on Ideas and Men, 1915–1916." *Public Administration* 48 (Spring 1970): 23–48.

McNeill, J. R. *Something New under the Sun: An Environmental History of the Twentieth-Century World.* New York: W. W. Norton, 2000.

Malchow, Howard LeRoy. *Agitators and Promoters in the Age of Gladstone and Disraeli: A Biographical Dictionary of the Leaders of British Pressure Groups Founded between 1865 and 1886.* New York: Garland, 1983.

———. "Public Gardens and Social Action in Late Victorian London." *Victorian Studies* 29 (Autumn 1985): 97–124.

Marsh, Arnold. "Octavia Hill and Clean Air: A Smoke Abatement Pioneer." *Society of Housing Managers Quarterly Journal* 5 (Oct. 1962): 21–23.

Marsh, Jan. *Back to the Land: The Pastoral Impulse in England from 1880 to 1914.* London: Quartet Books, 1982.

Mathias, Peter. *The First Industrial Nation: An Economic History of Britain, 1700–1914.* New York: Scribner, 1969.

Melosi, Martin V. "Cities, Technical Systems and the Environment." *Environmental History Review* 14 (Spring-Summer 1990): 45–64.

———. "The Place of the City in Environmental History." *Environmental History Review* 17 (Spring 1993): 1–23.

———. *The Sanitary City: Urban Infrastructure in America from Colonial Times to the Present.* Baltimore: Johns Hopkins University Press, 2000.

Messinger, Gary S. *Manchester in the Victorian Age: The Halfknown City.* Manchester: Manchester University Press, 1985.

Milan, Sarah. "Refracting the Gaselier: Understanding Victorian Responses to Domestic Gas Lighting." In *Domestic Space: Reading the Nineteenth-Century Interior,* edited by Inga Bryden and Janet Floyd, 84–102. Manchester: Manchester University Press, 1999.

Millward, Robert. "The Political Economy of Urban Utilities." In *The Cambridge Urban History of Britain.* Vol. 3, *1840–1950,* edited by Martin Daunton, 315–49. Cambridge: Cambridge University Press, 2000.

Mitchell, B. R. *British Historical Statistics.* Cambridge: Cambridge University Press, 1988.

———. *Economic Development of the British Coal Industry, 1800–1914.* Cambridge: Cambridge University Press, 1984.

———. *International Historical Statistics: The Americas, 1750–1993.* 4th ed. London: Macmillan Reference, 1998.

———. *International Historical Statistics: Europe, 1750–1993.* 4th ed. London: Macmillan Reference, 1998.

Mitchell, Sally, ed. *Victorian Britain: An Encyclopedia.* New York: Garland, 1988.

Mosley, Stephen. *The Chimney of the World: A History of Smoke Pollution in Victorian and Edwardian Manchester.* Cambridge: White Horse Press, 2001.

———. "Fresh Air and Foul: The Role of the Open Fireplace in Ventilating the British Home, 1837–1910." *Planning Perspectives* 18, no. 1 (2003): 1–21.

Murphy, Graham. *Founders of the National Trust.* London: Croom Helm, 1987.

Nead, Lynda. *Victorian Babylon: People, Streets, and Images in Nineteenth-Century London.* New Haven: Yale University Press, 2000.

Olsen, Donald. *The Growth of Victorian London.* New York: Holmes and Meier, 1976.

Parker, Albert. "Air Pollution from the Use of Fuels." In *Industrial Air Pollution Handbook,* edited by Albert Parker. London: McGraw-Hill, 1978.

———. "What's in the Air?" *Memoirs and Proceedings of the Manchester Literary and Philosophical Society* 111 (1968–69): 48–63.

Pearson, Kenneth, ed. *Drawn and Quartered: The World of the British Newspaper Cartoon, 1720–1970.* London: Times Newspapers, [1970].

Pelling, Margaret. *Cholera, Fever, and English Medicine, 1825–1865.* Oxford: Oxford University Press, 1978.

———. "Contagion/Germ Theory/Specificity." In *Companion Encyclopedia of the History of Medicine,* edited by W. F. Bynum and Roy Porter, 1:309–34. 2 vols. London: Routledge, 1993.

———. "'Progress, Difficulties, Suggestions, and Reforms': Public Health, 1888–1974." *Public Health* 102, no. 3 (1988): 209–15.

Pennybacker, Susan D. *A Vision for the Future, 1889–1914: Labour, Everyday Life, and the LCC Experiment.* London: Routledge, 1995.

Pick, Daniel. *Faces of Degeneration: A European Disorder, c. 1848–c. 1918.* Cambridge: Cambridge University Press, 1989.

Pickstone, John V. "Dearth, Dirt, and Fever Epidemics: Rewriting the History of British 'Public Health,' 1780–1850." In *Epidemics and Ideas: Essays in the Historical Perception of Pestilence,* edited by Terence Ranger and Paul Slack, 125–48. Cambridge: Cambridge University Press, 1992.

Porter, Dorothy. "'Enemies of the Race': Biologism, Environmentalism, and Public Health in Edwardian England." *Victorian Studies* 34 (Winter 1991): 159–78.

———. "Eugenics and the Sterilization Debate in Sweden and Britain before World War II." *Scandinavian Journal of History* 24, no. 2 (1999): 145–62.

Porter, Roy. *The Greatest Benefit to Mankind: A Medical History of Humanity.* New York: W. W. Norton, 1997.

Prochaska, F. "A Mother's Country: Mothers' Meetings and Family Welfare in Britain, 1850–1950." *History* 74 (Oct. 1989): 379–99.

Radice, Giles, and Lisanne Radice. *Will Thorne, Constructive Militant: A Study in New Unionism and New Politics.* London: George Allen and Unwin, 1974.

Ramm, Agatha. "Gladstone's Religion." *Historical Journal* 28, no. 2 (1985): 327–40.

Ranlett, John. "'Checking Nature's Desecration': Late-Victorian Environmental Organization." *Victorian Studies* 26 (Winter 1983): 197–222.

———. "The Smoke Abatement Exhibition of 1881." *History Today* 31 (Nov. 1981): 10–13.

Readman, Paul. "Landscape Preservation, 'Advertising Disfigurement,' and English National Identity c. 1890–1914." *Rural History* 12, no. 1 (2001): 61–83.

Rodgers, Daniel T. *Atlantic Crossings: Social Politics in a Progressive Age.* Cambridge: Belknap Press of Harvard University Press, 1998.

Rodhe, Henning, and Robert Charlson, eds. *The Legacy of Svante Arrhenius: Understanding the Greenhouse Effect.* Stockholm: Royal Swedish Academy of Sciences, 1998.

Rome, Adam W. "Coming to Terms with Pollution: The Language of Environmental Reform, 1865–1915." *Environmental History* 1 (July 1996): 6–28.

Rosen, Christine Meisner, and Christopher C. Sellers. "The Nature of the Firm: Towards an Ecocultural History of Business." *Business History Review* 73, no. 4 (1999): 577–600.

Rosen, Christine Meisner, and Joel A. Tarr. "The Importance of an Urban Perspective in Environmental History." *Journal of Urban History* 20 (May 1994): 299–310.

Rosen, George. *A History of Public Health.* Expanded ed. Introduction by Elizabeth Fee. Baltimore: Johns Hopkins University Press, 1993.

Rowbotham, Sheila, and Jeffrey Weeks, eds. *Socialism and the New Life: The Personal and Sexual Politics of Edward Carpenter and Havelock Ellis.* London: Pluto Press, 1977.

Samuel, Raphael. *Island Stories: Unravelling Britain.* Edited by Alison Light, with Sally Alexander and Gareth Stedman Jones. Vol. 2 of *Theatres of Memory.* London: Verso, 1998.

Scarrow, Howard A. "The Impact of British Domestic Air Pollution Legislation." *British Journal of Political Science* 2, no. 3 (1972): 261–282.

Schivelbusch, Wolfgang. *Disenchanted Night: The Industrialization of Light in the Nineteenth Century.* Translated by Angela Davies. Berkeley: University of California Press, 1988.

Searle, G. R. *Eugenics and Politics in Britain, 1900–1914.* Leyden: Noordhoff, 1976.

———. *The Quest for National Efficiency: A Study in British Politics and Political Thought, 1899–1914.* Oxford: Basil Blackwell, 1971.

Sellers, Christopher. "Factory as Environment: Industrial Hygiene, Professional Collaboration and the Modern Sciences of Pollution." *Environmental History Review* 18 (Spring 1994): 55–83.

Semmel, Bernard. *Imperialism and Social Reform: English Social-Imperial Thought, 1895–1914.* Cambridge: Harvard University Press, 1960.

Sheail, John. "Environmental History: A Challenge for the Local Historian." *Archives* 22 (1997): 157–69.

———. *An Environmental History of Twentieth-Century Britain.* New York: Palgrave, 2002.

Shifrin, Neil S., et al. "Chemistry, Toxicity, and Human Health Risk of Cyanide Compounds in Soils at Former Manufactured Gas Plant Sites." *Regulatory Toxicology and Pharmacology* 23, no. 2 (1996): 106–16.

Simmons, I. G. *An Environmental History of Great Britain: From 10,000 Years Ago to the Present.* Edinburgh: Edinburgh University Press, 2001.

Slack, Paul. *The Impact of Plague in Tudor and Stuart England.* Oxford: Clarendon Press, 1985.

Smith, Donald J. M. "The Eaton Hall Connection." *Transport History* 9, no. 3 (1978): 256–59.

Soloway, Richard. *Demography and Degeneration: Eugenics and the Declining Birthrate in Twentieth-Century Britain.* Chapel Hill: University of North Carolina Press, 1990.

Springhall, John O. "Lord Meath, Youth, and Empire." *Journal of Contemporary History* 5, no. 4 (1970): 97–111.

Stallybrass, Peter, and Allon White. *The Politics and Poetics of Transgression.* Ithaca: Cornell University Press, 1986.

Stange, G. Robert. "The Frightened Poets." In *The Victorian City: Images and Realities,* edited by H. J. Dyos and Michael Wolff, 2:475–94. London: Routledge and Kegan Paul, 1973.

Stedman Jones, Gareth. *Outcast London: A Study in the Relationship between Classes in Victorian Society.* 1971. Reprint, New York: Pantheon, 1984.

Steinberg, Theodore. *Acts of God: An Unnatural History of Natural Disaster in America.* New York: Oxford University Press, 2000.

Stepan, Nancy Leys. "Biological Degeneration: Races and Proper Places." In *Degeneration: The Dark Side of Progress,* edited by J. Edward Chamberlin and Sander L. Gilman, 97–120. New York: Columbia University Press, 1985.

———. *"The Hour of Eugenics": Race, Gender, and Nation in Latin America.* Ithaca: Cornell University Press, 1991.

———. *The Idea of Race in Science: Great Britain, 1800–1960.* London: Macmillan, 1982.

Stine, Jeffrey K., and Joel A. Tarr. "At the Intersection of Histories: Technology and the Environment." *Technology and Culture* 39 (Oct. 1998): 601–40.

Stradling, David. *Smokestacks and Progressives: Environmentalists, Engineers and Air Quality in America, 1881–1951.* Baltimore: Johns Hopkins University Press, 1999.

Stradling, David, and Peter Thorsheim. "The Smoke of Great Cities: British and American Efforts to Control Air Pollution, 1860–1914." *Environmental History* 4 (Jan. 1999): 6–31.

Stratton, Michael, and Barrie Stuart Trinder. *Twentieth Century Industrial Archaeology.* London: E. and F. N. Spon, 2000.

Supple, Barry. *The History of the British Coal Industry.* Vol. 4, *1913–1946: The Political Economy of Decline.* Oxford: Clarendon Press, 1987.

Susser, Mervyn, and Ezra Susser. "Choosing a Future for Epidemiology, I: Eras and Paradigms." *American Journal of Public Health* 86 (May 1996): 668–73.

Szreter, Simon. "The Importance of Social Intervention in Britain's Mortality Decline c. 1850–1914: A Re-Interpretation of the Role of Public Health." *Social History of Medicine* 1 (Apr. 1988): 1–37.

Tarn, J. N. *Five Per Cent Philanthropy: An Account of Housing in Urban Areas between 1840 and 1914.* London: Cambridge University Press, 1973.

Tarr, Joel A. *The Search for the Ultimate Sink: Urban Pollution in Historical Perspective.* Akron, Ohio: University of Akron Press, 1996.

———. "Searching for a 'Sink' for an Industrial Waste: Iron-Making Fuels and the Environment." *Environmental History Review* 18 (Spring 1994): 9–34.

———. "Transforming an Energy System: The Evolution of the Manufactured Gas Industry and the Transition to Natural Gas in the United States (1807–1954)." In *The Governance of Large Technical Systems,* edited by Olivier Coutard, 19–37. London: Routledge, 1999.

Taylor, Antony. "'Commons-Stealers,' 'Land-Grabbers,' and 'Jerry-Builders': Space, Popular Radicalism, and the Politics of Public Access in London, 1848–1880." *International Review of Social History* 40 (1995): 383–407.

Te Brake, W. H. "Air Pollution and Fuel Crises in Pre-Industrial London, 1250–1650." *Technology and Culture* 16, no. 3 (1975): 337–59.

Tesh, Sylvia Noble. "Miasma and 'Social Factors' in Disease Causality: Lessons from the Nineteenth Century." *Journal of Health Politics, Policy, and Law* 20 (Winter 1995): 1001–24.

Thomas, A. O., and J. N. Lester. "The Reclamation of Disused Gasworks Sites: New Solutions to an Old Problem." *Science of the Total Environment* 152 (1994): 239–60.

Thomas, Keith. *Man and the Natural World: A History of the Modern Sensibility.* New York: Pantheon, 1983.

Thorsheim, Peter. "Interpreting the London Fog Disaster of 1952." In *Smoke and Mirrors: The Politics and Culture of Air Pollution,* edited by E. Melanie DuPuis, 154–69. New York: New York University Press, 2004.

———. "The Paradox of Smokeless Fuels: Gas, Coke, and the Environment in Britain, 1813–1949." *Environment and History* 8, no. 4 (2002): 381–401.

Tuan, Yi-Fu. *Landscapes of Fear.* Minneapolis: University of Minnesota Press, 1979.

Uppenbrink, Julia. "Arrhenius and Global Warming." *Science* 272, no. 5265 (1996): 1122.

Waller, P. J. *Town, City, and Nation: England, 1850–1914.* Oxford: Oxford University Press, 1983.

Ward, David. "Victorian Cities: How Modern?" *Journal of Historical Geography* 1 (1975): 135–51.

Weindling, Paul. *Health, Race, and German Politics between National Unification and Nazism, 1870–1945.* Cambridge: Cambridge University Press, 1989.

Weinreb, Ben, and Christopher Hibbert, eds. *The London Encyclopædia.* New York: St. Martin's Press, 1983.

White, Lynn, Jr. "The Historical Roots of Our Ecologic Crisis." *Science* 155 (1967): 1203–7.

Williams, Raymond. *The Country and the City.* Oxford: Oxford University Press, 1973.

———. *Problems in Materialism and Culture.* London: Verso, 1980.

Williams, Trevor I. *A History of the British Gas Industry.* Oxford: Oxford University Press, 1981.

Wilson, J. F. *Lighting the Town: A Study of Management in the North West Gas Industry, 1805–1880.* London: P. Chapman, 1991.

Winslow, Charles-Edward Amory. *The Conquest of Epidemic Disease.* Princeton: Princeton University Press, 1943.

Winter, James. *Secure from Rash Assault: Sustaining the Victorian Environment.* Berkeley: University of California Press, 1999.

Wohl, Anthony S. *Endangered Lives: Public Health in Victorian Britain.* Cambridge: Harvard University Press, 1983.

Worboys, Michael. "From Miasmas to Germs: Malaria, 1850–1879." *Parassitologia* 36, nos. 1–2 (1994): 61–68.

Worster, Donald. *Nature's Economy: A History of Ecological Ideas,* 2nd ed. Cambridge: Cambridge University Press, 1994.

Wright, Patrick. *On Living in an Old Country: The National Past in Contemporary Britain.* London: Verso, 1985.

Wright, Peter, and Andrew Treacher, eds. *The Problem of Medical Knowledge: Examining the Social Construction of Medicine.* Edinburgh: Edinburgh University Press, 1982.

Yavetz, Ido. "A Victorian Thunderstorm: Lightning Protection and Technological Pessimism in the Nineteenth Century." In *Technology, Pessimism, and Postmodernism,* edited by Yaron Ezrahi, Everett Mendelsohn, and Howard Segal, 53–75. Dordrecht: Kluwer Academic Publishers, 1994.

Yeo, Eileen Janes. "Social Motherhood and the Sexual Communion of Labour in British Social Science, 1850–1950." *Women's History Review* 1, no. 1 (1992): 63–87.

Zuckerman, Arnold. "Disease and Ventilation in the Royal Navy: The Woodenship Years." *Eighteenth-Century Life* 11 (Nov. 1987): 77–89.

Index

Page numbers in *italic* indicate illustrations.

CPSIA information can be obtained
at www.ICGtesting.com
Printed in the USA
FFHW022027270119
50247836-55280FF